fragile lives

A heart surgeon's stories
of life and death on the
operating table

PROFESSOR
STEPHEN WESTABY

HarperCollins*Publishers*

HarperCollins*Publishers*
1 London Bridge Street
London SE1 9GF

www.harpercollins.co.uk

First published by HarperCollins*Publishers* 2017

5 7 9 10 8 6 4

© Stephen Westaby 2017
Illustrations © Dee McLean

Stephen Westaby asserts the moral right to
be identified as the author of this work

A catalogue record of this book is
available from the British Library

HB ISBN 978-0-00-819676-9
PB ISBN 978-0-00-820936-0

Printed and bound in Great Britain by
Clays Ltd, St Ives plc

MIX
Paper from
responsible sources
FSC˚ C007454

FSC™ is a non-profit international organisation established to promote
the responsible management of the world's forests. Products carrying the
FSC label are independently certified to assure consumers that they come
from forests that are managed to meet the social, economic and
ecological needs of present and future generations,
and other controlled sources.

Find out more about HarperCollins and the environment at
www.harpercollins.co.uk/green

This book is dedicated to my wonderful children
Gemma and Mark, and to my granddaughters
Alice and Chloe.

contents

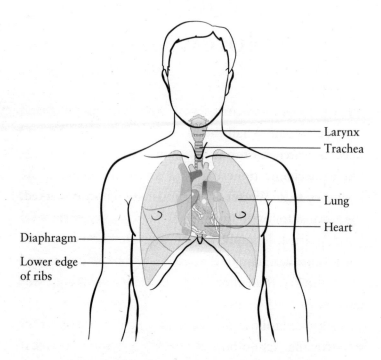

Position of the heart and the lungs in the chest

foreword

WOODY ALLEN FAMOUSLY SAID, 'The brain is my second favourite organ.' I had the same affinity with the heart. I liked to watch it, stop it, repair it and start it up again, like a mechanic tinkering with an engine beneath the bonnet of a car. When I finally understood how it worked, the rest just followed on. After all, in my younger days I'd been an artist. I simply shifted from brush on canvas to scalpel through human flesh. More hobby than job, and more pleasure than chore, it was simply something I was good at.

My career followed a curiously erratic course, from self-effacing schoolboy to wildly extrovert medical student, from ruthlessly ambitious young doctor to introverted surgical pioneer and teacher. Throughout this journey I was repeatedly asked what I found so compelling about cardiac surgery. I hope the following pages will make that clear.

But before launching into the action let me share with you some facts about this vibrant organ. Every heart is different. Some are fat, some are lean. Some are thick,

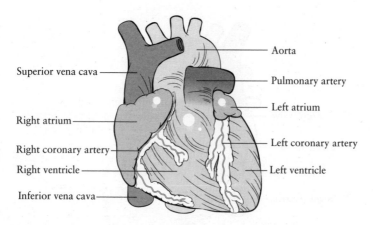

The heart viewed from the front

some are thin. Some are fast, some are slow. Just never the same. Most of the 12,000 that I've worked on have been desperately sick, causing misery, crushing chest pain, interminable fatigue and terrifying breathlessness.

What's so fascinating about the human heart is its movement – the rhythm and efficiency of the thing. The facts are staggering. The heart beats more than 60 times per minute to pump five litres of blood. This adds up to 3,600 beats an hour and 86,400 in 24 hours. It beats more than 31 million times in a year and 2.5 billion times in 80 years. The left and right sides of the heart eject more than 6,000 litres of blood daily to the body and lungs. A truly incredible workload that requires huge amounts of energy. So when the heart fails there are dire consequences. And given this astounding performance how could one conceive of replacing the human heart with a mechanical device? Or even with a dead person's heart?

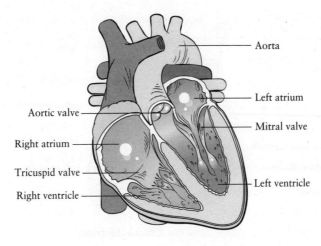

The heart chambers, valves and
major blood vessels from the front

My school biology classes taught me that the heart sits
in the centre of the chest and has four parts – two
collecting chambers, the left and right atria; and two
pumping chambers, the left and right ventricles. Textbook
diagrams show them side by side, like a house with two
bedrooms situated above a sitting room and a kitchen.
The spongy, expansible lungs surrounding the heart
resemble the roof of a Swiss chalet, and they constantly
replenish blood oxygen levels and expel carbon dioxide
into the atmosphere. (Most of us also know that other
chemicals can be discharged in the breath, notably alcohol
when blood levels exceed the liver's capacity to metabolise
it.)

Well-oxygenated blood leaves the lungs to enter the left
atrium through four separate veins, two from each side.

During the heart's filling phase, or diastole, blood flows through the mitral valve – named after its likeness to a bishop's mitre – and into the powerful left ventricle. During ventricular contraction, or systole, the mitral valve closes. The contents of the left ventricle are ejected onwards through the aortic valve into the aorta and around the arteries to the body.

Intriguingly, the right ventricle works in an entirely different way. It's crescentic in shape and applied to the side of the left ventricle, the part known as the ventricular septum. With this 'new moon' shape the right ventricle pumps like bellows. So the ventricles depend on each other. It was that rhythm of the heart that I found captivating, like watching a pianist's hands or a dancer's feet.

But is it all really that simple? My mother used to buy sheep hearts from the butcher; inexpensive and tasty enough, and great for dissecting. It was while cutting these up that I discovered that real hearts are more complex and difficult to understand than in textbook diagrams, as the shape and muscular architecture of the two ventricles are completely different. Nor are they left and right – more front and back. The thicker left ventricle is conical in shape and has circular muscle bands that constrict and rotate the chamber. Now we can visualise how the left ventricle really works. As the powerful muscle contracts and thickens, its cavity narrows and shortens. During relaxation – the diastolic phase – the left ventricle recoils and the aortic valve closes. The recoiling cavity widens and lengthens, sucking blood from the atrium into the ventricle through the mitral valve. Thus every coordinated cycle of contraction and relaxation involves narrowing, twisting and shortening, followed by widening, uncoiling and lengthening. A veritable Argentine tango ... but with two important differences: the whole process takes less than one second and the dance goes on forever.

Every cell in the body needs 'life blood' and oxygen; in the absence of these the tissues die at different rates, brain first, bone last. It all depends upon how much oxygen each cell needs. When the heart stops, the brain and nervous system are damaged in less than five minutes. Brain death ensues.

Now you are a cardiologist. You know about the heart and circulation. But you will still need a surgeon to help your patient.

1

the ether dome

*For this relief much thanks; 'tis bitter cold and
I am sick at heart.*

William Shakespeare, *Hamlet*, Act I, Scene i

THE FINEST OF MARGINS separates life from death, triumph from defeat, hope from despair – a few more dead muscle cells, a fraction more lactic acid in the blood, a little extra swelling of the brain. Grim Reaper perches on every surgeon's shoulder and death is always definitive. There are no second chances.

November 1966. I'm eighteen and a week into my first term as a student at the Charing Cross Hospital Medical School in the centre of London, just across from the hospital itself. I wanted to see a vibrant, beating heart, not a slimy piece of lifeless meat on the dissection table. I was told by a hall porter at the school that heart surgery was done over the road in the hospital on Wednesdays and that I should look for the ether dome. Find the green door on the very top floor under the eaves where nobody goes.

But don't get caught, he warned me. Pre-clinical students are not allowed up there.

It was late in the afternoon, already dark, and drizzle was falling on the Strand when I set out to find the ether dome, which turned out to be an antiquated leaden glass dome above the operating theatre in the old Charing Cross Hospital. Not since my entrance interview had I entered the hallowed portals of the hospital itself. We students had to earn that privilege by passing exams in anatomy, physiology and biochemistry. So I didn't enter by the Grecian portico of the main entrance, but sneaked in through Casualty under the blue light and found a lift, a rickety old cage used to take equipment and bodies from the wards to the basement.

I worried that I would be too late, that the operation would be finished – and that the green door would be locked. But it wasn't. Behind the green door was a dark, dusty corridor, a depository for obsolete anaesthetic machines and discarded surgical instruments. Ten yards away I could see the glow of the operating lights beneath the dome itself. It was an old operating theatre viewing gallery respectfully separated by glass from the drama on the operating table no more than ten feet below, with a hand rail and curved wooden benches worn smooth by the restless backsides of would-be surgeons.

I sat clutching the hand rail, just me and Grim Reaper, and peered through glass hazy with condensation. It was a heart operation and the chest was still open. I moved to find the best view, settling for a position directly above the surgeon's head. He was well known, at least in our medical school, a tall man, slim and imposing, with very long fingers. In the 1960s heart surgery was still new and exciting, its practitioners few and far between, and not many had been properly trained in the specialty. Often they were skilled general surgeons who had visited one of the pioneering centres and then volunteered to start a new programme. They were on a steep learning curve, with the cost measured in human lives.

The two surgical assistants and the scrub nurse were huddled together over the gaping wound, frantically shuttling instruments between them. And there it was, the focus of their attention and of my fascination. A beating human heart. In fact it was squirming more than beating, and was still attached by cannulas and tubing to the heart–lung machine. Cylindrical discs were spinning through a trough of blood bathed in oxygen and a crude roller pump squeezed the tubes, accelerating this life blood back to the body. I peered closer but could only see the heart as the patient was completely covered by green drapes, gratifyingly anonymous to all concerned.

The surgeon restlessly shifted his weight from foot to foot, wearing the big, white operating boots that surgeons once used to keep blood off their socks. The team had replaced the patient's mitral valve but the heart was struggling to separate from the bypass machine. This was the first time I'd seen a beating human heart, and even to me it looked feeble, blown up like a balloon, pulsating but not pumping. On the wall behind me was a box marked 'Intercom'. I threw the switch and the drama now had a soundtrack.

Against the din of amplified background noise I heard the surgeon say, 'Let's give it one last go. Increase the adrenaline. Ventilate and let's try to come off.'

There was silence as everyone watched the desperate organ fight for its life.

'There's air in the right coronary,' the first assistant said. 'Give me an air needle.' He shoved the needle into the aorta, frothy blood fizzed into the wound, then the patient's blood pressure started to improve.

Sensing a window of opportunity, the surgeon turned to the perfusionist. 'Come off now! This is our last chance.'

'Off bypass,' came the reply, said more as a blunt matter of fact than with any great confidence.

The heart–lung machine was switched off and the heart was now free-standing, with the left ventricle pumping blood to the body, the right ventricle to the lungs. Both were struggling. The anaesthetist stared hopefully at the screen, watching the blood pressure and heart rate. Knowing that this was their last attempt, the surgeons silently withdrew the cannulas from the heart and sewed up the holes, each one of them willing it to get stronger. For a while it fluttered feebly but then the pressure slowly drifted down. There was bleeding from somewhere – not torrential but persistent. Somewhere at the back. Somewhere inaccessible.

Lifting the heart caused it to fibrillate. It was now squirming again, wriggling like a bag of worms, but not contracting, fed by uncoordinated electrical activity. Wasted energy. It took the anaesthetist a while to spot this on his screen. 'VF,' he shouted. I'd soon learn that this meant ventricular fibrillation. 'Shock it.'

The surgeon was expecting this and was holding the defibrillating paddles hard against the heart. 'Thirty joules.' Zap! No change. 'Give it sixty.'

Zap! This time it defibrillated, but then just sat there stunned and devoid of electrical activity, like a wet brown paper bag. Asystole, as we call it.

Blood continued to fill the chest and the surgeon poked the heart with his finger. The ventricles responded by contracting. He poked it again and the rhythm returned. 'Too slow. Give me a syringe of adrenaline.' The needle was shoved unceremoniously through the right ventricle into the left, and a clear liquid squirted in. Then he massaged the heart with his long fingers to push the powerful stimulant into the coronary arteries.

The grateful heart muscle responded rapidly. Straight out of the textbook, the heart rate accelerated and the blood pressure soared, up and up, dangerously testing the stitches. Then, as if in slow motion, the cannula site in the aorta gave way. Whoosh! Like a geyser erupting, a crimson fountain hit the operating lights, spraying the surgeons and soaking the green drapes. Someone murmured, 'Oh, shit.' An understatement. The battle was lost.

Before a finger could plug the hole the heart was empty. Blood dripped from the lights and red rivulets streamed across the marble floor. Rubber soles stuck to it. The anaesthetist frantically squeezed bags of blood into the veins, but to no avail. Life was fast ebbing away. As the injected slug of adrenaline wore off, the turgid heart simply blew up like a balloon and stopped. Stopped forever.

The surgeons stood silently in despair, as they did week after week. The senior surgeon then walked away out of my view and the anaesthetist turned off the ventilator, waiting for the electrocardiogram to flatline. He removed the tube from the patient's windpipe, then he too disappeared from view. The brain was already dead.

Just yards away mist descended on the Strand. Commuters rushed into Charing Cross Station to get out of the rain, late lunches were finishing at Simpson's and Rules, cocktails were being shaken in the Waldorf and the Savoy. That was life, this was death. A lonely death on the operating table. No more pain, no more breathlessness, no more love, no more hate. No more anything.

The perfusionist wheeled his machine out of theatre, and it would take hours to disassemble, clean, restore and sterilise it for the next patient. Only the scrub nurse lingered. Then she was joined by the anaesthetic nurse who had comforted the patient in the anteroom. They took off their masks and stood silently for a while, unconcerned by the sticky blood that covered every surface and by the chest still splinted open. The anaesthetic nurse searched for the patient's hand beneath the drapes and held it. The scrub nurse pulled away the blood-soaked covering from the face and stroked it. I could see the patient was a young woman.

They were oblivious to the fact that I was upstairs in the ether dome. No one had seen me there. Only Grim Reaper – and he'd already departed with the soul. I gingerly shifted along the bench to look at the woman's face. Her eyes were wide open, staring up into the dome. She was ashen white but still beautiful, with her fine cheek bones and jet black hair.

Like the nurses I couldn't leave. I needed to know what happened next. They peeled back the bloodied drapes from her naked body. I was silently screaming for them to take out that hideous retractor cranking open her breast-bone and let her poor heart go back to where it belonged. When they did the ribs recoiled and the poor lifeless organ was covered again. It lay flat, empty and defeated in its own space, with just a fearsome, deep gash separating her swollen breasts.

The intercom was still switched on and the nurses started to talk.

'What'll happen to her baby?'

'Adopted, I guess. She wasn't married. Her parents were killed in the Blitz.'

'Where did she live?'

'Whitechapel, but I'm not sure the London do heart surgery yet. She got really sick during the pregnancy. Rheumatic fever. She nearly died during the delivery. Might've been for the best.'

'Where's the baby now?'

'On the ward, I think. Matron'll have to deal with it.'

'Does she know?'

'Not yet. You go and find her. I'll get some help to finish off.'

It was all so matter of fact. A young woman had died, her baby left without a relative in the world. No more love, no more warmth, lost amid that tangled, blood-soaked technology in the operating theatre. Was I ready for this? Was this what I aspired to?

Two student nurses came to wash the body. I recognised them as respectful public schoolgirls from the Friday-night freshers' dance. They'd brought a bucket of soapy water with sponges and set about scrubbing her clean. They removed the vascular cannulas and the bladder catheter but were visibly upset by the wound and what lay beneath. Blood kept slopping out of it.

'What did she have done?' asked the girl I'd danced with.

'Heart operation, obviously,' came the reply. 'Valve replacement, I guess. Poor kid. She's only our age. Bet her mum's upset.'

They covered the wound with gauze to soak the blood, then taped it up. The scrub nurse returned and thanked the girls for a job well done. She called back the surgical registrar to close over the wound, ready to move the body to the mortuary, as all deaths on the operating table are referred to the coroner for autopsy. The young woman would be sliced open again from neck to pubis, so there was no point closing the breastbone or bringing together the different layers of the chest wall. He took a big needle and some thick braid, and sewed her up like a mail bag. The wound edges still gaped and oozed serum. Mail bags were much neater.

It was now around 6.30 in the evening and I was meant to be in the pub down the road getting pissed with the rugby team. But I still couldn't leave. I was attached to this empty shell, this skinny corpse I'd never met but now felt I knew well. I'd been with her at the single most important part of her life.

The three nurses manhandled her into a starched white shroud with a ruff around the neck, tied it up at the back then secured her ankles with a bandage. She was beginning to stiffen with rigor mortis. The students had done their job with kindness and respect. I knew that I would meet them again. Maybe I'd ask them how they felt.

Now there were just the two of us left, the corpse and me. The operating lights still shone on her face and she was staring straight up at me. Why hadn't they closed her eyelids like they did in the movies? I could see through those dilated pupils to the pain etched on her brain.

From fragments of conversation I'd overheard and with just a little medical knowledge I could sketch her life story. She was in her twenties. Born in the East End. She could only have been a small child when her parents were killed in the bombing. As a child she carried the scars of those sights and sounds, the fear of being alone as her world disintegrated. Brought up in poverty, she develops rheumatic fever, a simple streptococcal sore throat that triggers a devastating inflammatory process. Rheumatic fever was common in areas of deprivation and over-crowding. Perhaps she had painful, swollen joints for a few weeks. What she doesn't know is that the same inflammation is in her heart valves. There was no diagnostic test in those days.

She develops chronic rheumatic heart disease and is known as a sickly child. Perhaps she develops rheumatic chorea – involuntary, jerky movements, unsteady gait and emotional turmoil. She gets pregnant, an occupational hazard. But this makes things worse as her sick heart must work much harder. She becomes breathless and swollen but makes it through to term. Maybe the London Hospital delivers her safely but recognises heart failure. A murmur. A leaking mitral valve. They prescribe the heart drug digoxin to make it beat stronger, but she doesn't take it as it makes her nauseous. Soon she's too tired and breathless to look after the baby, and she cannot lie flat. With worsening heart failure her outlook is grim. They send her to the city to see a surgeon, a real gentleman in a morning suit with pinstriped trousers. He's kind and sympathetic, and says that only surgery on her mitral valve can help. But it doesn't. It terminates her sad life and leaves another orphan in the East End.

When the porters came for her the operating lights had long been switched off. The mortuary trolley – a tin coffin on wheels – was drawn up alongside the operating table. By now her limbs were rigid. The body was unceremoniously dragged into this human sardine can, her head bouncing with a sickening thud, but nothing could hurt her any more. I was relieved to lose eye contact. A green woollen blanket was folded over the top to make it look like an ordinary trolley, and then off they went to slot her into the fridge. Her baby would never see her again, would never have a mother again.

Welcome to cardiac surgery.

* * *

I sat there, arms on the rail, chin on my hands, staring down from the ether dome at the black rubber surface of the empty operating table, as generations of would-be surgeons had done before me. The ether dome was a gladiatorial amphitheatre, people coming here to gaze down on a spectacle of life or death. Perhaps if others had been there it might have seemed less brutal, others with whom to share the shock of this poor girl's death, the misery awaiting her child.

Auxiliary nurses came with mops and buckets to erase the last traces of her – her blood now dry on the floor around the operating table, the bloody footprints heading towards the door, the blood on the anaesthetic machine, the blood on the operating lights. Blood everywhere – now meticulously wiped up. A slip of a girl reaching up to clean the operating light saw me in the dome, my pale face and staring eyes against the gloom. I frightened her, and so it was my cue to leave. But one spot of blood remained on top of the light where no one could see. Adherent and black, it said part of me is still here. Remember me.

The green door closed behind me and I walked away to the shuddering lift where her body had been taken down to lie in a cold fridge in the mortuary.

Notice of autopsies were posted on a board in the entrance hall of the medical school. Usually the patients were elderly. The young ones were either drug addicts, road-traffic accidents, suicides from the underground system or cardiac surgery patients. I found her on the list for Friday morning. She was called Beth. Not Elizabeth, just Beth. She was twenty-six years old. It had to be her. On the day of the autopsy the bodies were brought from the hospital mortuary in the basement, then dragged under the road to the medical school in a tin box on rails by a pulley system and up the lift to the autopsy room. Should I go? Should I watch her guts and brain be cut out, watch her dead heart be carved into slices, tell them how she really died in that crimson fountain?

No, I couldn't do it.

Beth taught me a very important lesson that day in the ether dome. Never get involved. Walk away as her surgeons did and try again tomorrow. Sir Russell Brock, the most renowned heart surgeon of the era, was known for his bluntness about losing patients – 'I have three patients on my operating list today. I wonder which one will survive.' This may seem insensitive, even callous, but to dwell on death was a dreadful mistake then, and it still is now. We must learn from failure and try to do better the next time. But to indulge in sorrow or regret brings unsustainable misery.

I grappled with this later in my career when my interests veered towards the sharp end: heart surgery for complex congenital anomalies in babies and young children. Some came toddling happily into the hospital, teddy bear in one hand, Mummy holding the other. Blue lips, little chest heaving, blood as thick as treacle. They'd never known a different life and I strived to provide that for them. To make them pink and energetic, liberate them from impending doom. I did this in good faith, yet sometimes without success. So what should I do? Sit with the weeping parents in a dark mortuary holding a cold, life-less hand, blaming myself for taking that risk?

All heart surgery is a risk. Those of us who make it as surgeons don't look back. We move on to the next patient, always expecting the outcome to be better, never doubting it.

2

humble beginnings

Courage is doing what you're afraid to do.
There can be no courage unless you're scared.

Edward V. Rickenbacker, *The New York Times*
Magazine, 24 November 1963

IT WAS AT THE VERY START of the post-war baby boom
that I arrived into the world in the maternity department
of Scunthorpe War Memorial Hospital on 27 July 1948,
star sign Leo. Good old Scunthorpe, my childhood home
for eighteen years, a steel town and the long-suffering butt
of music-hall jokes.

My dear mother, exhausted after a long and painful
labour but happy with her first child, brought me safely
back home from the carnage of the delivery suite. I was a
pink, robust son, wailing from the depths of his newly
expanded lungs.

My mother was an intelligent woman, caring, gentle
and well liked. During the war she'd managed a small
high-street bank, and with other tills empty the old folks

would still queue to tell her their troubles. My father joined the RAF at sixteen to fight the Germans, and after the war he got a job in the local Co-operative grocery department and worked hard to improve our circumstances. Life wasn't easy.

We were church-mice poor in a grimy council estate. House number 13, no pictures allowed on the walls in case the plaster crumbled, with a corrugated tin air-raid shelter in the back garden that housed geese and chickens – and the outside toilet.

My maternal grandparents lived directly across the street. Grandmother was kindly and protective of me, but frail. Grandfather worked at the steelworks and during the war had been the local air-raid warden. On pay day I'd go with him to the works to collect his wages. There I was intrigued by the spectacle of white-hot molten metal being poured into ingots, bare-chested, sweaty men in flat caps stoking the furnaces, steam trains belching fire, clanking up and down between the rolling mills and the slag heaps, and sparks flying everywhere.

Grandfather patiently taught me how to draw and paint. He'd sit over me, puffing away on Woodbines as I painted red night skies over the chimneys, street lamps and railway trains. Grandfather smoked twenty a day and spent his whole life working in smoke at the steelworks. Not the best recipe.

In 1955 we got our first television set, a 10-inch-square box with a grainy black-and-white picture and just one channel, the BBC. Television dramatically widened my awareness of the outside world. That year two Cambridge scientists, Crick and Watson, described the molecular structure of DNA. In Oxford the physician Richard Doll linked smoking with lung cancer. Then came exciting news on a programme called *Your Life in Their Hands* that would shape the rest of my life. Surgeons in the United States had closed a hole in the heart with a new machine. They called it the heart–lung machine, because it did the job of both organs. The television doctors wore long white coats down to the floor, the nurses had fine, starched uniforms and white caps and rarely spoke, and the patients sat stiffly to attention with their bed sheets folded back.

The show talked about heart operations and how surgeons at the Hammersmith Hospital would attempt one soon. They too would close holes in the heart. This seven-year-old street kid was captivated. Quite mesmerised. Right then I decided that I would be a heart surgeon.

At ten I passed the tests for entry to the local grammar school, and by then I was quiet, compliant and self-conscious. As one of the 'promising' set I was forced to work hard. I was a natural in art, although I had to stop those classes in favour of academic subjects. But one thing was clear. I was good with my hands, and my fingertips connected with my brain.

One afternoon after school I was out walking with Grandfather and his Highland terrier Whisky on the outskirts of town when he stopped dead on a hill, clutching the collar of his cloth shirt. His head bowed, his skin turned ashen grey and, sweating and breathless, he sank to the ground like a felled tree. He couldn't speak and I saw the fear in his eyes. I wanted to run and fetch the doctor but Grandfather wouldn't let me. He couldn't risk being off work, even at the age of fifty-eight. I just held his head until the pain abated. It lasted thirty minutes, and once he'd recovered we slowly made for home.

His ill health wasn't news to my mother. She told me that he'd been getting a lot of 'indigestion' while cycling to work. Reluctantly, Grandfather agreed to get off the bike, but it didn't do much good. The episodes became more frequent, even at rest, and especially when he climbed the stairs. Cold was bad for his chest, so the old iron bed was brought down in front of the fire and the commode was carried inside to save a journey outdoors.

His ankles and calves were so swollen with fluid that he needed bigger shoes. It was a gargantuan effort just to tie his shoelaces, and from then on he didn't get out much, mostly just moving from the bed to a chair in front of the fire. I'd sit and draw for him to take his mind off his rotten symptoms.

I remember that dismal wet afternoon in November, the day before President Kennedy was assassinated in Dallas. I came home from school to find a black Austin-Healey outside my grandparents' house. It was the doctor's car and I knew what that meant. I stared through the condensation on the front window but the curtains were drawn, so I went around the back of the house and walked in quietly through the kitchen door. I could hear sobbing and my heart sank.

The living-room door was ajar and inside it was dimly lit. I peered in. The doctor was standing by the bed with a syringe in his hand, and my mother and grandmother were at the end of the bed, clasping each other tightly. Grandfather looked leaden, with a heaving chest and his head tipped back, and frothy pink fluid was dripping from his blue lips and purple nose. He coughed agonally, spraying bloody foam over the sheets. Then his head fell to one side, wide eyes staring at the wall, fixed on the placard proclaiming 'Bless This House'. The doctor felt for a pulse at his wrist, then whispered, 'He's gone.' A sense of peace and relief descended on the room. The suffering was at an end.

The certificate would say 'Death from heart failure'. I slipped out unnoticed to sit with the chickens in the air-raid shelter, and quietly disintegrated.

Soon afterwards my grandmother was diagnosed with thyroid cancer, which started to close off her windpipe. 'Stridor' is the medical term to describe the sound of strangulation as the ribs and diaphragm struggle to force air through the narrowed airway, and that's what we heard. She went to Lincoln, forty miles away, for radiotherapy, but it burned her skin and made swallowing more difficult. We were given some hope of relief by an attempted surgical tracheostomy, but when the surgeon tried to do it he couldn't position the hole low enough in the windpipe below the narrowing. Our hopes were dashed and she was doomed to suffer until she died. It would have been better if they'd allowed her to go under anaesthetic. Every evening I sat with her after school and did what I could to make her comfortable. Soon opiate drugs and carbon dioxide narcosis clouded her consciousness, and one night she slipped away peacefully with a large brain haemorrhage. At sixty-three she was the longest-lived of my grandparents.

When I reached sixteen I took a job at the steelworks in the school holidays, but after a collision between a dumper truck and a diesel train hauling molten iron they dispensed with my services. I spotted a temporary portering job at the hospital and negotiated the role of operating theatre porter. There were disparate groups to please. The patients – fasted, fearful and lacking dignity in their theatre gowns – required kindness, reassurance and handling with respect. Junior nurses were friendly and fun, the nursing sisters were self-important, bossy and business-like, and needed me to shut up and do what they told me, and the anaesthetists didn't want to be kept waiting. The surgeons were simply arrogant and just ignored me – at first.

One of my jobs was to help transfer anaesthetised patients from their trolleys onto the operating table. I knew what sort of surgery was planned for each one, having read the operating list, and I helped out by adjusting the overhead lights, focusing them on the site of the incision (as an artist I was intrigued by anatomy and had some knowledge of what lay where). Gradually the surgeons began to take notice, some even asking me about my interest. I told them that I'd be a heart surgeon one day, and soon enough I was allowed to watch the operations.

Working nights was great because of the emergencies: broken bones, ruptured guts and bleeding aneurysms. Most of those with aneurysms died, the nurses cleaning up the corpses and putting on the shrouds, me hauling them from the operating table and onto the tin mortuary trolley, always with a dull thud. Then I'd wheel them off to the mortuary and stack the bodies in the cold store. I soon got used to it.

Inevitably my first mortuary visit took place in the dead of night. The windowless grey brick building was set apart from the main hospital and I was frankly terrified of what I'd find in there. I turned the key in the heavy wooden door that led directly into the autopsy room but when I reached inside I couldn't find the light switch. I'd been given a torch and its beam danced around as I plucked up the courage to go in.

Green plastic aprons, sharp instruments and shiny marble sparkled in the gloom. The room smelt of death, or what I expected death to smell like. Eventually the torch beam settled on a light switch and I turned on the overhead neons. They didn't make me feel any better. There were stacks of square metal doors from floor to ceiling – the cold store. I needed to find a fridge but wasn't sure which ones were empty.

Some doors had a piece of cardboard slotted into them with a name on it, and I figured that they must be occupied. I turned the handle on one without a name, but there was a naked old woman under a white linen sheet. An anonymous corpse. Shit. Try again on the second tier. This time I was lucky, and I pulled out the sliding tin tray and pushed the creaking mechanical hoist towards my stiff. How to make this thing work without dropping the body on the floor? Straps, crank handles and manhandling. I just got on with it and slid the tray back into the fridge.

The mortuary door was still wide open – I didn't want to be shut in there alone. I sped out and pushed the squeaking mortuary trolley back to the main hospital ready for the next customer. I wondered how pathologists could spend half of their career in that environment, carving entrails from the dead on marble slabs.

Eventually I charmed an elderly female pathologist into letting me watch the autopsies. Even after witnessing some disfiguring operations and terrible trauma cases this took some getting used to, young and old sliced open from throat to pubis, eviscerated, scalp incised from ear to ear and pulled forward over the face like orange peel. An oscillating saw removes the cranium, as if taking the top off a boiled egg, and then the whole human brain lies in front of me. How does this soft, grey, convoluted mass govern our whole lives? And how on earth could surgeons possibly operate on this, a wobbly jelly?

I learned so much in that dingy, desolate autopsy room: the complexity of human anatomy, the very fine line between life and death, the psychology of detachment. There was no room for sentiment in pathology. An ounce of compassion there may be, but affinity with the cadaver? No. Yet personally I felt sad for the young who came here. Babies, children and teenagers with cancer or deformed hearts, those whose lives were destined to be short and miserable or had been terminated by a tragic accident. Forget the heart as the source of love and devotion, or the brain as the seat of the soul. Just get on and slice them up.

Soon I could identify a coronary thrombosis, a myocardial infarction, a rheumatic heart valve and a dissected aorta, or cancer spread to the liver or lungs. The common stuff. Charred or decomposed bodies smelt bad, so Vicks ointment stuffed up the nostrils spared your olfactory nerves. I found suicides to be terribly sad, but when I verbalised this I was told to 'Get over it if you want to be a surgeon' and that it would all be easier when I was old enough to drink. I sensed that alcohol was high on the list of surgeons' recreational activities, and this seemed more obvious when they were called in at night. But who was I to judge?

I began to wonder whether I could really get in to medical school. I was no great academic, and I struggled with maths and physics. For me these subjects were the real barometer of intelligence. But I excelled in biology and could get by in chemistry, and in the end I passed a lot of exams, stuff I would never need like Latin and French literature, additional maths and religious studies. These I saw as a function of effort, not intelligence, but hard work bought me my ticket out of the council estate. In addition, the time spent in the hospital had made me worldly. I'd never been out of Scunthorpe, yet I knew about life and death.

I started to search for a place at medical school, and returned to the hospital during every school holiday. I progressed to working as an 'operating department assistant', becoming an expert in cleaning up blood, vomit, bone dust and shit. Humble beginnings.

I was surprised to be called for an interview at a magnificent Cambridge college. Someone must have put in a good word but I never learned who it was. The streets bustled with lively young students in their gowns chatting loudly with public school accents, all seeming much smarter than me. Erudite, bespectacled professors cycled down cobbled streets in their mortarboards off to college dinners for wine, then port. My mind flashed back to the grimy steelworkers silently making their way home in flat caps and mufflers through the smog for bread and potatoes, and then maybe a glass of stout. My spirits started to sink. I didn't belong here.

The interview was conducted by two distinguished fellows in an oak-panelled study overlooking the main college quadrangle. We sat in well-worn leather armchairs. It was meant to be a relaxed atmosphere, and nothing was said about my background. The anticipated question, 'Why do you want to study medicine?' never came. Wasted interview practice. Instead I was asked why the Americans had just invaded Vietnam and whether I had heard of any tropical diseases their soldiers might be exposed to. I didn't know whether there was malaria in Vietnam so I said, 'Syphilis.'

That broke the ice, particularly when I suggested that this might be less of a health problem than napalm and bullets. Next I was asked whether smoking cigars may have contributed to Winston Churchill's demise (he'd only recently died). Smoking was one of the key words I was waiting for. My mouth fired off in automatic mode: cancer, bronchitis, coronary artery disease, myocardial infarction, heart failure, how the corpses of smokers looked in the autopsy room. 'Had I seen an autopsy?' 'Many.' And cleared up the brains, guts and bodily fluids afterwards. 'Thank you. We'll let you know in a few weeks.'

Next I was called down to Charing Cross Hospital, between Trafalgar Square and Covent Garden on the Strand. The original hospital was built to serve the poor of Central London and had a distinguished war history. Although I arrived early I was always last alphabetically, so I twiddled my thumbs anxiously to while away what seemed like hours. A kindly matron received the candidates with tea and cakes, and I made polite conversation with her about what had happened to the hospital during the war.

The interview took place in the hospital board room. Across the other side of the boardroom table from me was the chief interviewer – a distinguished Harley Street surgeon wearing a morning suit – together with the famously irascible Scottish professor of anatomy upon whom the *Doctor in the House* films were based. I sat straight-backed to attention on an upright wooden chair – no slouching here. I was first asked what I knew about the hospital. Thank you, God. Or Matron. Or both. Next I was asked about my cricketing record and whether I could play rugby. And that was all, the interview was over. I was the last of the day, they'd had enough and they'd let me know.

I wandered out into Covent Garden past the colourful market stalls and bristling public houses. All life was there: tramps, tarts, buskers and bankers, the Charing Cross Hospital clientele, and the black cabs and scarlet London buses that drove up and down the Strand. Meandering between the crowds and the traffic I came to the grand entrance of the Savoy Hotel. I wondered whether I dared go in. Surely I looked smart enough in my interview suit and Brylcreemed hair. But the decision was swiftly made for me when the immaculate doorman pushed the swing doors open and ushered me through with a 'Welcome, sir.' The seal of approval. From Scunthorpe to the Savoy.

I strode purposefully through the atrium, past the Savoy Grill, hesitating only to scrutinise the menu in its gilt frame. The prices! I didn't stop. A sign pointed to the American Bar. The hall was lined with signed cartoons, photographs and paintings of West End stars, and when I reached it there was no queue as it was only 5 pm. Perched on a high stool I furtively devoured free canapés and perused the cocktail menu. Devoid of insight – this was my first alcoholic drink – I was pushed to make a decision. 'Singapore Sling, please.' Like flipping a switch, my life had changed. Had I ordered a second I'd never have found King's Cross station.

Within the week a letter arrived from Charing Cross Hospital Medical School. Opening it surrounded by my anxious parents felt like defusing a bomb. There was the offer of a place. The conditions? Just pass my biology, chemistry and physics exams, no grades specified. Charing Cross was a small medical school with an intake of only fifty students each year, but I'd be following in the footsteps of distinguished alumni such as Thomas Huxley the zoologist and David Livingstone the explorer. I was the first in my family to go to university, the first to attempt to become a doctor and, hopefully, the first heart surgeon.

3

lord brock's boots

*He has been a doctor a year now and has had
two patients. No, three, I think. Yes, three.
I attended their funerals.*

Mark Twain

THE BEST WAY TO PREPARE for the exams to become a
Fellow of the Royal College of Surgeons was to work as
an anatomy demonstrator in the dissection room of the
medical school, teaching anatomy to the new students in
minute detail and helping them to dismantle their cadaver
sliver by sliver – skin, fat, muscle, sinew and then the
organs. They were given greasy embalmed corpses on a
tin trolley, and there were six new and impressionable
students to each one. They'd march in with their starched
white coats and brand new dissection kits – scalpel, scissors, forceps and hooks in a linen roll – all as green as
grass. Just like me when I started.

I moved from group to group to maintain their momentum. A few couldn't hack it. Spending untold hours pick-

ing away at a corpse was not part of their medical dream, so I gave the best advice I could to help them through it: wear strong perfume, don't skip breakfast and try to think about something else – football, shopping, sex, anything. Just learn enough to pass the tests and don't let the stiffs drive you out. This worked with some. Others had nightmares, their dissected corpses visiting them at night.

For my first surgery exam I had to master anatomy, physiology and pathology – nothing to do with being able to operate. There were courses in London that just hammered home the facts, taught by past examiners who presented the information in the way that the college wanted it. Pay up and pass was the message, unless you were an idiot. Yet two-thirds of candidates still failed come exam time, including myself on the first occasion.

In the midst of this academic monotony the Royal Brompton Hospital advertised for 'Resident Surgical Officers', with Fellowship of the Royal College of Surgeons being 'desirable but not obligatory'. Could I aspire to this? I'd only just passed the first part. It would be a minimum of three years before I could sit the final exam, but there would be nothing lost by trying for the post.

Despite the odds I succeeded in securing the job and started in the position just a few weeks later. I was allocated to work for Mr Matthias Paneth, an imposing six-foot, six-inch German, and Mr Christopher Lincoln, the newly appointed children's heart surgeon of similar height. Two very different personalities, but both scary in their own way until I knew them better. In my massively busy junior resident jobs at Charing Cross I learned that the only way to keep up was to write everything down. Record every order or request as it was verbalised. To forget was to be in deep shit, so I always carried a clipboard. This was a source of great amusement to Mr Paneth, who took to saying, 'Did you get that, Westaby? Did you get that, Westaby?'

My surgical logbook opened in spectacular fashion. The Paneth team had a case scheduled after the outpatient's clinic, a little old lady from Wales for mitral valve replacement. The boss invited me to go and start while he saw a couple more private patients. I proudly changed into the blue scrubs. Not only that, I found a pair of white rubber surgeon's boots in an open locker. They were well worn and dirty. I could have had new clogs but coveted these discarded second-hand boots. Why? Because down the strip at the back was written 'Brock'. I was about to inherit Lord Brock's boots.

By now Baron Brock of Wimbledon was seventy and had stopped operating, Paneth alluding to his having 'perpetual disappointment at the unattainability of universal perfection'. He was President of the Royal College of Surgeons when I was at medical school and stayed on as Director of the Department of Surgical Sciences, and now I'd be following in his footsteps. Literally. I strode out of the surgeons' changing room straight into the operating theatre to introduce myself.

The old lady was on the operating table. The scrub sister, who had already prepared her with antiseptic iodine solution and covered her naked body in faded green linen drapes, was now impatiently tapping her theatre clogs on the marble floor, and the long-suffering anaesthetist Dr English and the chief perfusionist were playing chess by the anaesthetic machine. I sensed that everyone had been waiting for some time. I pulled on my face mask and quickly scrubbed up, relishing this first opportunity to showcase my skills.

I carefully located the landmarks, the sternal notch at the base of the neck and the tongue of cartilage at the lower end of the breastbone. The scalpel incision – a perfectly straight line cut from top to bottom – would carefully join the two. The old lady was thin and emaciated with heart failure, and there was little fat between skin and bone to cleave with the electrocautery. At this point there was still no sign of the other assistant surgeon, but I pressed on regardless, seeking to impress the nurses.

I took the oscillating bone saw and tested it. Bzzzz. That was fierce enough. So I bravely started to run it up the bone towards the neck. Then, disaster. After the light spattering of bloody bone marrow there was a sudden whoosh of dark red blood pouring from the middle of the incision. Oh shit! Instantly I started to sweat, but Sister knew the score, swiftly moving around to the first assistant's position. I grabbed the sucker but she was giving the orders. 'Press hard on the bleeding.'

Dr English belatedly looked up from the chess board, unfazed by the frenetic activity. 'Get me a unit of blood,' he calmly instructed the anaesthetic nurse. 'Then give Mr Paneth a call in Outpatients.'

I knew what the problem was. The saw had lacerated the right ventricle. But how? There should have been a tissue space behind the sternum and fluid in the sac around the heart. Sister was reading my mind, something she would do many times over the next six months. 'You do know that this is a reoperation.' A statement that was really a question.

'No, absolutely not,' I replied frantically. 'Where's the bloody scar?'

'It was a closed mitral valvotomy. The scar's around the side of the chest. You can just see it under her breast. Didn't Mr Paneth tell you it was a re-do?'

By this point I'd decided to keep my mouth shut. It was time for action, not recrimination.

In reoperations the heart and surrounding tissues are stuck together by inflammatory adhesions, and there's no space between the heart and the fibrous sac around it. In this case the right ventricle had stuck to the underside of the breastbone and everything was matted together. Worse still, the right ventricle was dilated because the pressure in the pulmonary artery was high, the rheumatic mitral valve having narrowed considerably. We were there to replace the diseased valve but I'd buggered it up right from the start. Great.

Pressing hadn't controlled the bleeding. Blood still poured through the bone and the sternum wasn't completely open yet. The patient's blood pressure began to sag and, as she was a small lady, she didn't have that much blood to lose. Dr English started to transfuse donor blood but that wasn't the answer, like pouring water into a drainpipe. In one end, straight out the other. I was the surgeon, it was my job to stop the haemorrhage – and for that I needed to see the hole.

My own perspiration dripped into the wound and trickled down my legs into Lord Brock's boots. The old lady's blood flowed off the drapes onto the faded white rubber. By now one of the circulating nurses had scrubbed up and joined us at the operating table. Not so brave now, I lifted the saw again and asked Sister to move her hands. Through a deluge of blood I ran the saw through the remaining intact bone – the thickest part of the sternum, just below the neck. Then we pressed on the bleeding again while more transfusion restored some blood pressure.

As pressure drops the rate of bleeding slows. This gave me a window of opportunity to dissect the heart sufficiently away from the back of the breastbone to insert the metal sternal retractor and wedge open the chest. Now I could see the lacerated right ventricle spewing its contents into the wound. When everything is stuck together like this, spreading the bone edges can tear the heart muscle wide open, sometimes irretrievably. But I'd been lucky and her heart was still in one piece. Just about.

By now my own pulse was galloping. I could see that the problem was a ragged slit 5 cm long in the free wall of the right ventricle, comfortably distant from the main coronary arteries. Sister instinctively put her fist directly on it as I wound the retractor open, and this at last stemmed the bleeding. Dr English squeezed a second unit of blood in through the drips, bringing the old lady's blood pressure back up to 80 mm Hg, and the back-up scrub nurse divided the long plastic tubes to the heart–lung machine so that we could use it when ready. But as yet not enough of the heart had been exposed for that. First I needed to stitch up the bloody hole. As a surgical houseman I'd stitched skin, blood vessels and guts – never a heart.

Sister told me what stitch to use, and that it was best to stitch over and over rather than using individual stitches. This was quicker and would provide a better seal. 'Don't tie the knots too tight,' she added, 'or the stitches will cut through the muscle. She's fragile. Get started and you might finish before Paneth gets here and chews your head off.'

The difficult part was to stitch accurately as blood poured out of the ventricle with every beat. By now my gloves were dripping with blood on the outside and sweat on the inside, and sewing was all but impossible.

Dr English saw this and shouted, 'Use the fibrillator! Stop the heart beating for a couple of minutes.'

The fibrillator is an electrical device that causes what we'd normally never want to see – ventricular fibrillation, where the heart doesn't pump but quivers, stopping blood flow to the brain at normal body temperature. In four minutes brain damage begins.

Dr English was reassuring. 'Just defibrillate it after two minutes. If you haven't closed it by then we can wait a couple of minutes, then fibrillate again.'

I felt like a puppet with the experienced players pulling the strings. That was fine by me, so I put the fibrillating electrodes on the surface of what muscle I could see and Dr English threw the switch. The heart stopped beating and started quivering, and I began to sew at top speed. Just then Mr Paneth appeared at the operating theatre door. He could see ventricular fibrillation on the monitor and feared the worst. But I didn't look up and just kept on stitching. By the time Dr English announced the two-minute cut-off I'd almost finished bringing the muscle edges together. I carried on to three minutes. Then the hole was closed, with just the knot to tie.

Putting the defibrillating paddles as close to the heart as possible I said, 'Defibrillate.' Nothing happened. The leads to the paddles hadn't been plugged into the machine, a minor detail. Seconds ticked by. Then came the 'zap' I'd been waiting for. The heart briefly stood still then fibrillated again.

Paneth strode across from the door in his smart suit and outdoor shoes. No hat, no mask. He looked over the drapes at the quivering muscle and said the obvious. 'More volts.' Another zap. The heart defibrillated and started to beat vigorously.

Paneth grinned, then asked, 'Anything you'd like to tell me, Westaby? The mitral valve isn't in the right ventricle, you know. I thought you were bright.' He winked at Sister, announced that he was going for tea and meanwhile not to let Westaby do anything stupid.

I scraped my nerves from the ceiling, took stock and tied that last knot. The heart seemed to be working fine, despite my assault. There was blood all down my gown, on Lord Brock's boots and in a pool on the marble floor, but the blood pressure was back to normal. Today's battle had been won.

I looked at Sister, who was just a pair of cool blue eyes above the mask, and reached for her blood-stained rubber glove to say thanks for saving both of us. By the time Mr Paneth took over it was as if nothing had happened, apart from jokes about the extra needlework on the front of the heart. I felt like screaming at him, 'Why didn't you tell me she was a fucking re-do?', then realised that he probably had no recollection of that as it was many months since he'd talked to her in Outpatients.

The rest of the operation went smoothly. Dr English and the perfusionist continued their chess game, I held the sucker and Paneth chopped out the deformed valve, replacing it with a 'ball in cage' prosthesis. Then lots of stitching-up.

There was no end to the day for surgical residents. That night I sat in the intensive care unit waiting for the old lady to wake up, desperately hoping that she wasn't brain damaged and wondering how I'd have felt had she bled to her death on the operating theatre floor. Would I have had the grit to continue? Or would my surgical career have ended that day? There was such a very fine line between hero and zero, but I'd survived. I just wanted her to wake up now.

Her husband and daughter were keeping vigil by her bedside. Her husband asked whether the operation had gone well. I just glibly said, 'Yes, very well. Mr Paneth did a great job,' avoiding any implication that I'd fucked up.

As if to order, she opened her eyes. A wave of relief flowed over me. Husband and daughter jumped to their feet, making sure that she could see them as she stared up at the ceiling, still transfixed by the breathing tube. They reached out for her hand. At that point I realised something – heart surgery might become an everyday occurrence for me, but for the patient and their relatives it is once in a lifetime, and absolutely terrifying. Treat them kindly.

Cardiac surgery is like quicksand. Once in it you're sucked deeper and deeper, and I struggled to leave the hospital in case something remarkable happened and I missed it. I spent endless hours sitting beside the cots of Mr Lincoln's babies, listening to the bip, bip, bip of the monitors, watching the blood pressure sag and trying to get it up again, hoping that blood would stop dripping into the drains.

The next débâcle followed quite quickly. One Saturday evening before Christmas, a group of residents were in the pub following dinner in the mess. Because there was no casualty department at the Brompton it was highly unusual for emergency operations to be held at night, particularly over the weekend. With a couple of pints of beer on board we were alerted by the switchboard that an American Air Force jet had taken off from Iceland carrying a young man injured in a road-traffic accident. He had a tear in the wall of the aorta and Mr Paneth was coming in to operate. Bad problem, both the injury and the beer. Not so much the amount of alcohol – we were used to that – more the volume of urine to pass during a four-hour operation. Nor could I avoid being involved, as Paneth would need two assistants. Although there was no

way I could maintain concentration with a bursting bladder, I didn't want to lose face by asking to leave, like a
whimpering schoolboy with his hand up in class.

As the senior registrar went off to make arrangements
with the operating theatres I pondered the possibilities.
What about a urinary catheter and drainage bag for the
duration of the procedure? I didn't really relish the idea
of passing the catheter myself. Nor the discomfort of
standing with the bag of urine strapped to my leg. And
then it dawned on me. Lord Brock's operating theatre
boots! One of them would hold a couple of pints, and
with a length of Paul's tubing – thin-walled rubber tubing
that was once used for incontinent males – there would be
less risk of a bladder infection than if I inserted my own
urinary catheter.

I went to the wards in search of the tubing. This came
in a roll to be cut to the appropriate length, in my case
that of my inside leg. Once I'd found a supply, off I went
to the surgeons' changing room as I was keen to be in
theatre all ready to go – with my clipboard and white
boots as usual, tubing attached with sticky tape – when
the boss arrived. And I was just in time, the ambulance
screeching in from Heathrow much sooner than we'd
anticipated. Those jets were fast.

We were opening through the ribs of the left side of the
chest by midnight and soon encountered bleeding. Paneth
was in an irascible mood, having been called out of a
Christmas party. As I predicted the beer soon began to
make its effects tell and my registrar colleague became
restless, shifting from foot to foot and losing concentra-

tion. Eventually he had to excuse himself, so I moved into the first assistant position, coughing loudly to disguise the unusual squelching sound. I stayed in his position after he returned as I had no discomfort, despite the fact that my right Wellington boot was slowly filling. After another twenty minutes the registrar had to go out again.

By now the patient was safe, but Paneth was cross. 'What's wrong with him? He's been in the pub, hasn't he? He's been drinking.'

'I really don't know about that, Mr Paneth. I've been studying in the library all evening,' I replied, waiting to be struck down by a thunderbolt. But it never came.

'Well done, Westaby,' he said instead. 'You get on and close the chest. He can assist you for a change. See you on Monday.'

I disposed of the evidence and accompanied the young man back to intensive care. No one ever knew.

Now beyond sleep, I sat drinking coffee in the paediatric intensive care unit. I talked with the nurses while watching tiny people struggle for life at Christmas in their cosy incubators. As surgical trainees we were all chronically sleep deprived, but there was little excitement in sleep. Sleep was something for the odd weekend off. We were adrenaline junkies living on a continuous high, craving action. From bleeding patients to cardiac arrests. From theatre to intensive care. From pub to party.

Sleep deprivation underpins the psychopathy of the surgical mind – immunity to stress, an ability to take risks, the loss of empathy. Bit by bit I was joining that exclusive club.

4

township boy

Genius is one per cent inspiration,
ninety-nine per cent perspiration.

Thomas Edison

OCTOBER 1979. I was Senior Registrar with the thoracic surgery team at Harefield Hospital in north London. Everyone training in heart surgery had to learn to operate on the lungs and gullet as well, and this meant working with cancer, which I found deeply depressing. Too often it had already spread to other parts of the body, and for most patients the prognosis was grim, so they were depressed too. Moreover, there was an element of monotony about it. The choices were stark: between taking out half a lung or the whole lung, on the right or on the left, or removing the upper part of the gullet or the lower half. After doing each one of these procedures a hundred times it was no longer very stimulating.

Every so often a more challenging case would present itself. Mario was a forty-two-year-old Italian engineer working on a restoration project in Saudi Arabia. A jovial family man, he'd gone to the kingdom hoping to earn enough money to buy a house, which meant toiling hours on end at a large industrial complex outside Jeddah in the searing desert heat. Then catastrophe. Without warning, while he was working in an enclosed area, a huge boiler exploded, filling the air with steam. Steam under high pressure. It scalded his face and burnt the lining of his windpipe and bronchial tubes.

The shock almost killed him immediately. The scalded tissues were dead and whole sheets of necrotic membrane sloughed off from the lining of his bronchial tubes. This obstructive debris had to be removed through an old-fashioned rigid bronchoscope, a long brass tube with a light on its end passed through the back of his throat and voice box then down into his airways.

Mario needed this done regularly, almost daily, to prevent asphyxiation, and pushing the bronchoscope back and forth through his larynx became more and more difficult. Soon it became so scarred that the bronchoscope would not pass and he needed a tracheostomy – a surgical hole in the neck to enable him to breathe. But the dead bronchial lining was quickly replaced by inflammatory tissue and masses of cells started to fill the airways like calcium blocking water pipes. He became unable to breathe, and his condition took a relentless downhill course.

I took the call from Jeddah. The burns doctor looking after him explained the dire situation and wondered whether we had any advice. My only suggestion was that they airlift him to Heathrow and we'd see if anything could be done, so the building company paid for the medical evacuation and he arrived the following day. At the time my boss was in the twilight of his career and was happy for me to take on as much as I felt confident to do. Which was everything. I had no fear. But this was a disaster in a middle-aged man. I asked that we should take a look down his windpipe together and then try to come up with a plan.

Mario was a sorry sight. He was gasping for breath, with the infected froth pouring from his tracheostomy tube making a dreadful, gurgling sound. His scarlet face was badly burnt, its crusted, dead skin peeling away and weeping serum. Burnt on the outside and burnt on the inside, the fragile and bloody tissue that occluded the whole of his windpipe was going to asphyxiate him. It was a great relief for him to be put to sleep.

As he lapsed into unconsciousness I sucked blood-stained sticky secretions from the hole in his neck, then attached the tubing from the ventilator to the tracheostomy tube and squeezed the black rubber bag. The lungs were difficult to inflate against the resistance. I decided that we should attempt to pass the rigid bronchoscope by the normal route directly through the vocal cords and larynx. This is akin to sword swallowing, but down the airways rather than the gullet.

We needed a view of the whole windpipe and both right and left main bronchial tubes. For this the head needs to be tipped at the correct angle so the vocal cords at the back of the throat can be seen. We do try hard not to knock out any teeth. This technique used to be performed on conscious individuals after lung surgery, when I'd have to hoover the patients out because there were never enough physiotherapists. Rough at the time but better than drowning.

I manoeuvred the rigid telescope over the teeth and along the back of the tongue, then peered down to locate the snippet of cartilage – the epiglottis – that protects the opening of the voice box during swallowing. If you lift its tip with the bronchoscope you should be able to find the glistening white vocal cords, with a vertical slit between the two. This is the way into the windpipe and I'd done the procedure hundreds of times to biopsy lung cancer. Or remove peanuts. But here, with the voice box burned and the vocal cords like sausages, inflamed and angry looking, there was no way through. Mario was entirely reliant on the tracheostomy.

Standing aside, I tried to show the boss by keeping the bronchoscope still, propped on the teeth. He grunted and shook his head. 'Try pushing it harder. Nothing to lose, I suspect.'

Taking aim again, I pushed the beak of the scope where the slit should be and shoved. The swollen vocal cords parted and the instrument crashed against the tracheostomy tube. We attached the ventilating apparatus to the side of the bronchoscope and pulled out the tube. Normally we'd see the full length of the windpipe to where it divides into the main bronchi. In this case, not a chance. The airways had been virtually obliterated by the proliferating cells, so I eased the rigid implement onward using the sucker to aspirate blood and detached tissue, at the same time pushing in oxygen through the broncho-scope tip. I was hoping to see an end to the burns, and we finally encountered normal airway lining halfway down each main bronchial tube. But now the traumatised lining was oozing blood.

Mario's bright red face had turned purple and was getting bluer by the minute, so the boss took over, peering down the tube, occasionally inserting the long telescope for a better view. It was a precarious situation without an obvious solution. If you can't breathe you die. Fortunately with time the bleeding died down and the airway was better than it had been once some gunk had been removed. We reinserted the tracheostomy tube and put him back on the ventilator. Both sides of the chest still moved and both lungs were inflated. This was a triumph in itself, but it was doubtful there was any way forward. We both concluded that his prospects were bleak.

Two days later Mario's left lung collapsed and we went through the same process again. It was just as bad. The tissue just kept on growing, and he remained fully conscious on the ventilator but very distressed.

Asphyxiation is the most miserable way to die. I remembered my grandmother, strangulated by cancer of the thyroid gland. She'd been told she needed a tracheos-tomy, only to have the procedure aborted, so she sat propped up in bed day and night gasping for breath. I recalled trying to work out ways to help. Why wasn't it possible to put a tube further down, past the obstruction? Why couldn't tracheostomy tubes be made longer? A simple concept but I was repeatedly told it wasn't possible.

From what I could see through the bronchoscope, the situation with Mario was nearly identical. He needed something to bypass his whole trachea and both main bronchi, otherwise he'd be dead in days. We couldn't keep opening the airways with a bronchoscope. Not forever. Grim Reaper was winning this battle and was about to swing his scythe.

Ever the optimist, I questioned whether there was anything else we could do. Could we make a branched tube to bypass the damaged airways? The boss thought not, because it would clog with secretions. Surely someone else would have done it before for cancer. Then something else occurred to me – a company called Hood Laboratories in Boston, Massachusetts made a silicone rubber tube with a tracheostomy side limb, called a Montgomery T-tube after its ear, nose and throat surgeon inventor. Maybe I should talk to them and explain the problem.

When I bronchoscoped Mario later that afternoon I took measurements to calculate how long the tube needed to be to reach down each main bronchus, and in the evening I rang Hood. A small family firm who were most helpful, they confirmed that no one had tried such an approach but agreed to make me the bifurcated tube to fit the whole of Mario's trachea and main bronchial tubes. I said we needed it urgently. They delivered in less than one week, with no invoice, pleased to help with this unique case. Now I had to work out how to get it in.

I'd need to railroad the branched end of the tube into the separate bronchi simultaneously over guide wires. But wires were too sharp and dangerous for the delicate silicone rubber, and I needed something blunt and harmless to do the job. We used to dilate strictures of the gullet with gum elastic bougies. Two of the narrowest bougies would fit down the T-Y tube, and down each limb of the Y branches. I could insert the bougies through the damaged trachea and into one bronchus at a time, then railroad the tube into place over them. I drew the technique step by step and showed it to the other thoracic surgeons. The consensus was that we had absolutely nothing to lose. Without some crazy new approach Mario was definitely going to die.

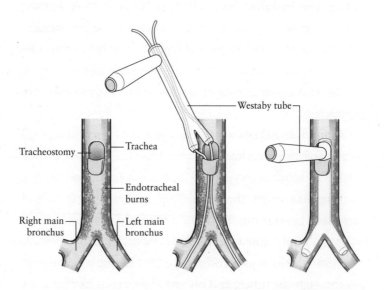

Technique for the insertion of the Westaby tube

The following day we took him to theatre, removed the tracheostomy tube and inserted the rigid bronchoscope through his burnt larynx. I tried to create as little bleeding as possible this time. We surgically enlarged the tracheostomy hole through which the T-Y tube would be introduced, then the bougies were inserted into the right and left main bronchi under direct vision through the scope, vigorously ventilating with 100 per cent oxygen between each step. So far so good. I lubricated the silicone rubber with K-Y jelly and shoved the tube forcefully downward. The bronchial limbs spread out at the branching point until there was resistance to any further pushing. It was in. Better than sex. The boss took a leap of faith and withdrew the bronchoscope into the larynx.

Ever the Irishman he exclaimed, 'Crikey, look at this! You're a bloody genius, Westaby.' The horribly disintegrating trachea had been replaced by a clean white silicone tube, the Y limbs sitting in perfect position. There was no kinking or compression, and clean healthy airways lay beyond.

Meanwhile Mario was blue and hypoxic. We were all so excited that we had stopped ventilating him, so we needed to blow in oxygen furiously. But he was now easy to inflate through the wide rubber airways. It was a complete revelation. Whether it would last we didn't know, and only time would tell. It depended entirely upon whether Mario was strong enough to cough secretions out through the tubes, and on our ability to suck them out and ventilate him through the side limb. When the swelling in his larynx and vocal cords subsided we'd keep this closed with the rubber bung. Then he could breathe and speak through his own larynx if it ever recovered. There were many unknowns, but for now Mario was safe. He could breathe. Fifteen minutes later he woke up with fantastic symptomatic relief.

I should have been thrilled that the concept had worked but I wasn't. I was in a difficult head space. I had a beautiful baby daughter – Gemma – whom I didn't live with. I lived at the hospital. This was grinding away at me in the background, so I compensated by operating fanatically on everything that I could lay my hands on. I was always available but was possessed by a disquieting restlessness.

In the meantime Mario recovered well, though life was difficult without a voice. He could cough secretions through the tube and keep it clear – something that everyone else had regarded as impossible – and was sent home to his family in Italy. Gratifyingly Hood started to manufacture the T-Y stent and called it the 'Westaby tube'. We used it often for patients in whom lung cancer was threatening to occlude their lower airways, relieving the dreadful strangulation that my grandmother was forced to endure. Why could no one have done it when she needed help and I was so miserable?

I never knew how many Westaby tubes were manufactured but it stayed in Hood's product list for many years. My original drawings were published in a chest surgery journal and served as the guide for others. While I still performed thoracic surgery I continued to use it for complex airways problems, often on a temporary basis until radiotherapy or cancer drugs caused the tumour to shrink. It was my grandmother's legacy. Then came the rare opportunity to use the artificial airways alongside my expertise with the heart–lung machine.

In 1992 I was invited to Cape Town for a conference to celebrate the twenty-fifth anniversary of the world's first heart transplant by Christiaan Barnard. At that meeting the distinguished children's heart surgeon Susan Vosloo asked me to see a sick two-year-old who'd been a patient at the Red Cross Children's Hospital for several weeks. Little Oslin lived in a sprawling Cape Town ghetto situated between the airport and the city, acre upon acre of tin shacks, wooden sheds and tents with brackish water and little sanitation. Nevertheless he was a cheerful little chap whose toys were oil drums, tin cans and pieces of wood. He knew no other life.

One day his family's faulty gas cylinder exploded in the shack, setting fire to the walls and roof. The blast killed Oslin's father outright, while Oslin sustained severe burns to his face and chest. Worse still, he inhaled hot gas from the blast, much like Mario. The accident department at Red Cross saved his life, intubating and ventilating him before he asphyxiated, then treating his burns with intravenous fluids and antibiotics. The little lad could survive the external burns, but his burnt-out trachea and main bronchi were life threatening, and without repeated bronchoscopies to clear slough and secretions he was destined to asphyxiate. On top of this his face was badly disfigured, he was almost blind and he couldn't swallow food, just his own saliva. So he was fed with liquids through a tube directly into the stomach.

It so happened that Susan had read a journal article about Mario and the tube I'd designed, and, although Oslin was much smaller, she wondered whether we could do anything to help him. When I first met the lad he was wearing a bright red shirt, had tight, curly black hair, and was pushing himself around the ward on a kiddies' bicycle with his back to me. Susan called to him and he turned around. The sight of his face took my breath away. There was no hair on the front of his scalp and no eyelids, just white sclera and a severely burned nose and lips. His neck was webbed from contracting scars with a tracheostomy tube in the middle. And the noise coming from him was heart-rending, a kind of rattling with thick mucus secretions made up by a long, noisy in-drawing of breath then a high-pitched wheeze as he forcibly exhaled. It was worse than a horror movie and tragic beyond belief. My first thought was, 'Poor kid, he should have died with his dad. It would have been much kinder.'

Strangely enough he was happy, as he'd never had a bicycle before the explosion. I kneeled on the floor to talk with him. He looked straight at me but I couldn't tell whether he could see my face as his corneas were opaque, so I took his little hand. There would be no objectivity in this discussion. I needed to help him, even if I wasn't sure how it could be done. We could work that out.

By this point I was chief of cardiac surgery in Oxford and I had to get back there to operate. In any case there was no Westaby tube in Cape Town, and if there had been it wouldn't have fitted anyway since the adult size was too big. Could I persuade Hood over in Boston to make a smaller tube? Probably, but not within the time frame that we'd been presented with; if he developed pneumonia in the next couple of weeks he'd surely die.

My return flight to Heathrow was the following day, so instead of going for lunch in the harbour I asked Susan whether she'd take me to see Oslin's township. Cape Town was my favourite city in the world but this was an aspect I'd never seen before, the sort of place that warranted an armed escort through its thousands of acres of misery and depravity. I'd come back in a couple of weeks when I had the tube, and a surgical strategy – that's what flying time was for. I quickly had it clear in my mind and before the plane touched down in Heathrow I'd drawn up the operation in detail.

I was back at the Children's Hospital in three weeks. There had already been a fund-raising drive to help Oslin and they expected to pay my expenses. But none of that mattered. I was driven to help the boy as no kid on earth deserved that. I guess thousands of Vietnamese children suffered the same with napalm, but I hadn't met them. I did know Oslin and I cared about him. So did the doctors and nurses at Red Cross. Perhaps the whole of Cape Town cared. As the airport taxi reached the city I saw the newspaper billboards emblazoned with 'UK Doc flies in to save dying Township boy' stuck on lamp-post after lamp-post. No pressure then.

At the hospital I met Oslin's mother for the first time. She'd been at work when the gas cylinder exploded and was now clearly depressed. She said virtually nothing, but signed the consent form for an operation that even I didn't understand.

We operated the following morning. I'd needed to trim the adult tube by shortening both bronchial limbs, the tracheostomy T-piece and the top part that would sit below his vocal cords, but even this shortened adult tube wouldn't fit inside the two-year-old's scarred windpipe. My objective was to rebuild his major airways around the tube. If it worked he'd have even wider airways than before the accident.

Clearly he wouldn't be able to breathe or be ventilated during the reconstructive surgery, so we'd do it with him supported on the heart–lung machine. This meant we'd open his sternum as we would in a heart operation. The tricky part was to gain access to the whole length of the trachea and main bronchial tubes from an incision in the front of his chest, these structures being situated directly behind the heart and large blood vessels.

I'd already worked it all out on a cadaver in the dissecting room in Oxford. When a sling was placed around the aorta and the adjacent vena cava they could be pulled apart to expose the back of the pericardial sac, like opening a pair of curtains and looking out onto a tree. Then a vertical incision between the two served to expose the lower trachea and both main bronchi.

My plan was to fillet these damaged tubes then lay in the modified T-Y stent. Next we'd repair the front of the opened airways and cover the tube with a patch of Oslin's own pericardium. It would be just like sewing an elbow patch onto a worn jacket sleeve. Simple. It should all heal up around the tube and we could maybe remove the prosthesis in time, after the tissues had healed and moulded around the silicone. That was my plan, in any case. Maybe 'fantasy' would have been a more realistic term, but no one else had a better solution.

The skin incision started in Oslin's neck just below his voice box and extended all the way down to the cartilage at the lower end of his breastbone. Since he was emaciated, unable to eat, there was no fat, so the electrocautery cut straight through to the bone, which we then sawed through. I cut out his fleshy, redundant thymus gland and dissected down onto the upper part of his inflamed trachea, all while he was ventilated through his tracheostomy tube. We needed to go on bypass before removing this and exposing the rest of his airways. The metal retractor stretched open his scarred little chest, exposing more of the fibrous pericardium. The front of this was removed for the tracheal patch and I saw that his little heart was beating away happily. Rarely do I see a normal child's heart, as most are deformed and struggling.

When I was ready to open the windpipe we started the bypass machine. This rendered the lungs redundant so we could remove the contaminated tracheostomy tube from the clean surgical field. Through the hole the devastation was clear to see. Poor Oslin had been breathing through a sewer. I cut down the length of it with the electrocautery and continued the incision into each main bronchus until I could see normal respiratory lining just at the limits of our access. Copious thick secretions poured out of the obstructed airways, then we scraped tissue off the walls, which caused all-too-predicable bleeding.

But the electrocautery eventually stopped the haemorrhaging, so we inserted the shiny white T-Y tube and covered it with a patch of Oslin's own pericardium. I adjusted the length of the rubber cylinder for the last time to get it just right, then sewed the patch into place to seal the implant. It needed to be airtight, otherwise the ventilator would push air into the tissues of the neck and chest, making him blow up like the Michelin man. With the shiny new breathing tubes attached to the ventilator we blew air into his little lungs. There was no leak. Both inflated then deflated normally. A sense of excitement permeated the room. The high-risk strategy was working.

Oslin's heart bounced off the bypass machine and his lungs moved freely, needing much lower pressure from the ventilator. Our anaesthetist murmured, 'Unbelievable. I'd never have believed it possible.' I covered the repair by closing the back wall of the pericardium, then asked that the registrar put in the drains and close.

Through the theatre window we could see Oslin's mother sitting in the waiting room, still expressionless and rigid with fear. I anticipated a blunt response to our news. But she was too emotionally drained to register relief, simply holding out her hand and squeezing mine. She whispered, 'God bless you,' then a tear zigzagged down her pockmarked cheek. I wished her a better life in the future, one way or another.

The intensive care unit was pleased to have him back. Most of their patients were township kids having heart surgery, and some of the nurses lived in that same environment. They'd cared for Oslin and his depressed mum for weeks, watching them both deteriorate. So 'UK Doc' had flown in to save 'Township boy' and succeeded. I was proud of that. Now it was time to ride off into the sunset.

Oslin recovered and could breathe freely through the white rubber tube in his neck. He couldn't speak but went on to have his corneal transplants. Being able to breathe and see at the same time was as much as he could have hoped for. The little family were relocated to better social housing on the outskirts of the city – crude but clean, and safer. A chest infection could still kill him, so for the first few months following the operation I contacted Cape Town frequently. Oslin was doing fine and Mum was faring better on anti-depressives. Then I stopped calling.

Eighteen months passed, and then a letter arrived from the Red Cross Hospital. Oslin had been found dead at home and no one really knew why. Sometimes life is shit.

the girl with no name

Dream that my little baby came to life again,
that it had only been cold, and that we rubbed
it before the fire and it had lived. Awake
and find no baby.

Mary Shelley, author of *Frankenstein*

THE GIRL WAS HAUNTINGLY BEAUTIFUL, with eyes that burned like lasers – as if the blistering desert heat were not enough (50°C during the day). When she fixed those eyes on mine she delivered a message – eye to eye, pupil to pupil, retina to retina – straight into my cerebral cortex. As she stood there holding her bundle of rags I understood perfectly what she was saying: 'Please save my child.' But she never spoke. Not to any of us. Ever. And we never even knew her name.

The Kingdom of Saudi Arabia, 1987. I was young and fearless, seemingly invincible and massively overconfident, and had just been appointed as a consultant in

Oxford. So why was I in the desert? Heart operations cost money. We'd worked hard to build Oxford's new cardiac centre and clear a backlog of sick hearts, but the annual budget was gone in five months so the management closed us down. Bugger the patients. The cardiologists were told to send them to London again.

On the day before I was locked out of the operating theatre I took a call from a prestigious Saudi cardiac centre that served the whole Arab world. Their lead surgeon needed three months sick leave, and they were looking for a locum who could tackle both congenital and adult heart surgery, an extremely rare species. At the time I wasn't interested but the following day I was, and three days later I jumped on the plane.

It was *Jumada al-thani*, the 'second month of dryness' in the Middle East, and I'd never felt heat like it, blistering, unremitting heat with the hot *shamal* wind blowing sand into the city. But it was a great cardiac centre. My medical colleagues were an eclectic mix of Saudi men who had trained overseas, Americans rotating from the major centres for experience, then the band of mercenaries from Europe and Australasia.

Nursing was very different. Saudi women did not nurse, as the profession was regarded with suspicion and disrespect, and was culturally taboo because it required mixing with the opposite sex. So all female nurses were foreign, most with contracts for just one or two years. Their accommodation was free, they paid no tax and stayed just long enough to save for that elusive mortgage back home. In turn they were not allowed to drive, had to travel in the rear of buses and be completely covered in public.

I was intrigued by my new environment: the repetitive calls to prayer from the minarets, the tantalising aromas of sandalwood, incense and amber around the hospital, Arabian coffee roasting on the frying pan or boiling with cardamom. It was a very different life and important not to step out of line – their culture, their rules, harsh penalties.

This presented a unique opportunity for me as I could operate on every conceivable congenital anomaly. There were innumerable young patients with rheumatic heart disease sent from remote towns and villages, mostly without access to anticoagulant therapy or drugs that we take for granted in the West. The rural health care was out of the Middle Ages, and we had to innovate and improvise to repair their heart valves rather than replace them with prosthetic materials. I remember thinking that every cardiac surgeon should train here.

One morning a bright young paediatric cardiologist from the Mayo Clinic, the world-famous medical centre in Minnesota, came to find me in the operating theatre. His opening gambit was, 'Can I show you something really interesting? Bet you haven't seen anything like this before,' swiftly followed by, 'Sadly, I doubt you can do anything about it.' I was determined to prove him wrong even before I'd seen the case because for surgeons the unusual is always a challenge.

He thrust the X-ray onto a light box. On a plain chest X-ray the heart is simply a grey shadow, but to the educated eye it can still tell the story. The message was clear. This was a small child with an enlarged heart in the wrong side of the chest, a rare anomaly called dextrocardia. Normal hearts lie to the left. In addition there was fluid on the lungs. But dextrocardia alone does not cause heart failure. There had to be another problem.

The enthusiastic Mayo cardiologist was testing me. He had already catheterised the eighteen-month-old boy and knew the answer. I offered an insightful guess to show off – 'In this part of the world it could be Lutembacher's syndrome.' This is a dextrocardia heart with a large hole between the right and left atrium, together with rheumatic fever that narrows the mitral valve, a rare combination which floods the lungs with blood, leaving the rest of the body short. The Mayo man was impressed. But no cigar!

He then wanted to take me to the catheterisation laboratory to see the angiogram (moving X-ray pictures with dye shot into the circulation to clarify the anatomy). By now I'd become fed up with the quiz but I still went along with him. There was a huge, sinister mass within the cavity of the left ventricle below the aortic valve, almost cutting off the flow of blood around the body. I could see this was a tumour, and whether benign or malignant the infant could not survive for much longer. So could I remove it?

I'd never seen surgery on a dextrocardia heart before. Few young surgeons had and most never would, but I did know about heart tumours in children. Indeed I'd published a paper on the subject in the United States that the paediatric cardiologist had read, making me the expert on the subject in Saudi Arabia.

The most common tumour in babies is a benign mass of abnormal heart muscle and fibrous tissue called rhab-domyoma. This is often associated with a brain abnormality that causes epileptic fits. No one knew whether the poor boy had suffered fits, but he was certainly dying from an obstructed heart. I asked the boy's age and whether his parents understood the desperate nature of the condition. Then his tragic story began to unfold.

It happened that the boy and his young mother were close to death when the Red Cross found them on the border between Oman and South Yemen. In the searing heat both were emaciated, dehydrated and in a state of collapse. Apparently she'd carried her son through the desert and mountains of Yemen, frantically seeking medical help. They were airlifted to the Military Hospital in Muscat in Oman, where they'd found that she was still trying to breastfeed. She'd nothing else to feed her son but her milk had dried up. When the boy was rehydrated with fluids into a vein he became breathless and was diagnosed with heart failure. In turn the mother had severe abdominal pain and a high temperature from a pelvic infection.

Yemen was a lawless place. She'd been raped, abused and mutilated. Not only that, she was African, not an Arab. The Red Cross suspected that she'd been kidnapped from Somalia and taken across the Gulf of Aden to be sold as a slave. But for one curious reason they couldn't be sure. She never spoke. Not a word. And she barely showed any emotion, even in response to pain.

When the Omanis saw the boy's chest X-ray and diagnosed dextrocardia and heart failure they transferred him to our hospital. Now, the Mayo man wondered whether I could conjure up a miracle. I knew that the Mayo Clinic had a great children's heart surgeon so I tentatively asked my colleague what Dr Danielson would do.

'Operate, I guess,' he said. 'Not a lot to lose, as it's all downhill from here.' That's what I expected him to say.

'Right then, I'll do what I can,' I said. 'At least we'll know what kind of tumour it is.'

What else did I need to know about the boy? Not only was his heart in the wrong side of the chest, but the abdominal organs were switched over too. What we call situs inversus. So the liver was in the left upper quadrant of the belly with the stomach and the spleen on the right. The bigger problem was that there was a large hole between the left atrium and right atrium, so blood returning from both the body and the veins of the lungs mixed freely. This meant that the level of oxygen in the arteries to his body was lower than it should be. Had his skin not been black he may have been recognised as a blue baby, where blood in the veins streams across into the arteries. Complicated stuff, even for doctors.

Money was no object here. We had state-of-the-art echocardiography, which in those days was new and exciting. It employed the same ultrasound waves that were used to detect submarines underwater, and an accomplished operator could provide sharp pictures of the inside of the heart and measure pressure gradients across areas of obstruction. I saw a clear image of the tumour in the small left ventricle, smooth and round, like a bantam's egg, and felt confident that it was benign. If only I could relieve him of it, the tumour would not grow back.

My plan was to clear the obstruction and close the hole in the heart, an ambitious attempt to restore normal physiology. This was straightforward in principle yet taxing in a back-to-front heart in the wrong side of the chest, and I didn't want any surprises. So I did what I always do in difficult circumstances – set about to draw the anatomy in detail.

Was the surgery possible? I didn't know, but we had to try. Even if we failed to remove the tumour completely it would still help him, although should it prove to be a rare malignancy his outlook would be bleak. But between us we were convinced that it was a rhabdomyoma.

It was time to meet the boy and his mother. Mayo man took me to the paediatric high-dependency unit where he was still being fed via a tube through his nose, which he disliked intensely. His mother was sitting cross-legged on a mat on the floor beside the cot and she never left his side, day or night.

As we approached she rose up. She was not at all what I'd expected – stunningly beautiful, with a striking resemblance to the model Iman, the widow of David Bowie. Her jet black hair was straight and long, her skinny arms folded across her chest. The Red Cross had established that she was Somali, and as she was a Christian her head was not covered.

Her long delicate fingers were clutching the bundle in which she held her son, precious rags that had shielded him from the hot sun then kept him warm in the cold desert nights. An umbilical cord of drip tubing emerged from these swaddling clothes and stretched to the drip pole and a bottle of feed, which was a milky-white solution replete with glucose, amino acids, vitamins and minerals to put meat back on his little bones.

Her eyes turned towards the stranger, the English heart surgeon whom she had heard about. Head gently tilted backwards in an attempt to remain detached, a bead of sweat appeared in the root of her neck and slithered down over the sternal notch. She was becoming anxious and her adrenaline was flowing.

I tried to engage with her in Arabic. '*Sabah al-khair, aysh ismuk?*' (Good morning, what's your name?). She said nothing and looked at the floor. Showing off, I continued, '*Terref arabi?*' (Do you know Arabic?), then, '*Inta min weyn?*' (Where are you from?). Still no response. Finally in desperation I asked, '*Titakellem ingleezi?*' (Do you speak English?). '*Ana min ingliterra*' (I'm from England).

Then she looked up, wide eyed, and I knew that she understood. Her lips parted but still no words. She was mute. Mayo man was speechless too, stunned by my linguistic skills, which unbeknownst to him had almost been exhausted. She appeared to appreciate my efforts and her shoulders dropped. She relaxed. I wanted to show her kindness, to take her hand, but I couldn't in this environment.

I indicated that I'd like to examine the boy, which was fine as long as she could continue to hold him. But I was shocked as she pulled back his linen covering. The lad was emaciated, with all his skinny ribs protruding. There was virtually no fat on him and I could see his bizarre heart pounding against his chest wall. He was breathing rapidly to overcome the stiffness in his lungs, his protuberant belly full of fluid and his enlarged liver clearly visible on the opposite side to normal. From the different skin tone to his mother I assumed that his father was an Arab. A curious rash covered his dark olive skin and I thought I saw fear in his eyes.

His mother pulled the linens back over his face, protectively. He was all she had in the world, this boy and a few rags and rings, and I couldn't help the upwelling of pity I felt for both of them. Surgery was my business but I was sucked into this whirlpool of despair, my objectivity gone.

In those days I had a red stethoscope and I placed it on the infant's chest, trying to look professional. There was a harsh murmur as blood squeezed past the tumour and out through the aortic valve, then the crackling sounds of wet lungs, even the gurgling and bubbling of empty guts. The cacophony of the human body.

Next I said, 'Mumken asaduq?' ('Will you let me help you'?). For a second I thought she responded. Her lips moved and those eyes fixed on mine. I sensed that she'd murmured, 'Naam' (Yes). I tried to explain that I needed to operate on the boy's heart to make him well so they could both have a better life. When tears appeared in her eyes I knew that she understood.

But how could I persuade her to sign a consent form? We sent for a Somali interpreter who repeated my words, yet still we had nothing in return. She remained impassive as I struggled to convey the complexities of the surgery. Name of the operation: 'Relief of left ventricular outflow tract obstruction in dextrocardia'. Then a short sentence for my own benefit: 'High risk case!' Absolving me of any blame, on paper at least. I was confident that this was the boy's only chance of survival, so just an 'X' from the mother would be enough. But she was signing over her whole life, her only reason for living. Eventually she took the pen from my hand and scribbled on the form, then I asked Mayo man to countersign and I signed it myself, looking directly into her eyes not at the document, searching for approval, I guess. By now her skin was glistening with perspiration; she was pouring out adrenaline and trembling with anxiety.

It was time for us to leave her alone. I indicated that I'd do the operation on Sunday when the best paediatric anaesthetist was available, then I said goodbye in both English and Arabic to show that I was still making an effort.

This was Thursday afternoon, the day before the Saudi weekend, and my colleagues were planning to take me out to the desert to camp on the dunes beneath the night sky as a way of escaping the oppression of the city. The convoy left in the early evening, just as the searing heat was starting to abate. When we ran out of road, the jeeps ploughed through the sand, miles and miles of it. They had a rule – never travel in just one vehicle. If it broke down that could be the end, even within twenty miles of the hospital.

The desert night was clear and cold. We sat around the campfire drinking homemade hooch and watching shooting stars. A Bedouin camel train passed by silently barely two hundred yards away, swords and Kalashnikovs glinting in the moonlight. They didn't even acknowledge us.

I felt uneasy and wondered how the mother had coped. Walking at night, hoping to find shelter during the day, and carrying water and child together, she must have been fuelled by hope but little else. However difficult it would prove to be, I was driven to save the boy and watch them both grow stronger.

The operation would be far from straightforward as I was still unsure of how to get at this tumour. The obstruction could only be accessible by opening the apex of the left ventricle wide, and that would impair its pumping ability. I kept working through the steps of the operation in my mind, always returning to the same question – 'What if?' With conventional surgery the technical challenges posed by this dextrocardia heart were virtually insurmountable. So would the boy be better off if he were operated on by a more experienced surgeon in the States? I couldn't see why, because his combination of pathologies was probably unique. No one else would have greater experience, even if they did have a better team. I had a good enough team and great equipment, the best that money could buy. So I was the man for the job, wasn't I?

It was then that I had my eureka moment, while staring up at the Milky Way. I suddenly knew the obvious way to get at the tumour. It might have been an outrageous idea, but I had a plan.

On Saturday I brought the anaesthetic and surgical teams together to discuss the case, and showed them the novel pictures of the unusual anatomy. Then, unusually – as much of what happens in an operating theatre remains impersonal, which is perhaps best when operating on those who may not survive – I told them the heart-rending story of the mother and boy. Everyone agreed that the boy was doomed if we made no effort but voiced justifiable concerns that the tumour was inoperable in dextrocardia. I said that we'd only know that through trying, although I kept the operating plan to myself.

I spent a hot, restless night in the apartment, my mind racing, disturbed by irrational thoughts. Would I have risked this back in England? And was I doing it for the patient or for the mother – or even for myself, so I could publish a paper about it? If I succeeded, who would care for this slave girl and her illegitimate child? The boy was an inconvenience. In Yemen he would be left out under a bush for the wolves to eat. It was the mother they wanted.

The early-morning call to prayer put an end to my discomfort. It was already 28°C as I walked from the apartment to the hospital. Mother and boy came down to the operating theatre complex and anaesthetic room at 7 am. She'd stayed awake until morning with her child in her arms, and all through the night the nurses had been concerned that she might capitulate and run away. She didn't, but they were still worried whether she would hand the boy over.

Despite premedication he was screaming and thrashing around when they tried to put him asleep. Dreadful for

the mother – and difficult for the anaesthetic staff – this
was pretty much routine in paediatric surgery. Gas
through the face mask eventually subdued him sufficiently
to insert a cannula into a vein and stun him into unconsciousness. His mother still wanted to follow him into the
operating theatre, so the ward nurses eventually dragged
her away. Finally raw emotion had broken through the
mask – whatever she had suffered physically, this was
worse for her. Yet there were still no words.

I sat, dispassionate, in the coffee room until the mayhem
abated, enjoying thick Turkish coffee and dates for breakfast. The caffeine hit was good for my ADHD but racked
up my sense of responsibility. What if the boy dies? Then
she has nothing. Nobody in the world.

One of the Australian scrub nurses came through to
ask that I check the equipment, the extra bits I'd requested
for the radical plan conceived under the dark desert sky.
I'd yet to share it with my team.

Uncovered on the shiny black vinyl of an operating
table, this emaciated little body was a pathetic sight, with
none of the puppy fat that every infant deserves. Instead
his skinny legs were swollen with fluid. The heart failure
paradox – the muscle is replaced by water but the weight
stays the same. His prominent, skinny ribs rose and fell
with the ventilator, as he was no longer struggling for
breath on his own. Now everyone understood why the
mother was so fiercely protective. We could see the heart
beating away in the wrong side of the chest and the
outline of his swollen liver in the contrary side of the
bulging abdomen. Everything was the wrong way round,

all a source of fascination for the onlookers and present-
ing a daunting challenge for me. I'd seen one operation on
dextrocardia in the US and another at Great Ormond
Street. This would be the first I'd attempted myself.

There were still streaks of dried salt down his cheeks
from the traumatic separation from his mother. What was
it I used to say when asked if I was ever anxious about
undertaking an operation? 'No. It's not me on the table!'
But although I don't do anxiety, I was now in tiger country with an untested procedure in an unfamiliar environment and could feel sweat trickling down my back. It all
felt a very long way from Oxford.

Everyone was happier when that fragile little body
was covered in blue drapes, leaving just a rectangular
window of dark skin exposed over the breastbone. He
was now no longer a child, just a surgical challenge. That
is until we heard his tormented mother banging on the
operating theatre doors. She'd given her minders the slip
and rushed back, and after a brief struggle they allowed
her to sit in the corridor just outside. Her day had been
traumatic enough without being dragged away for a
second time.

Back inside the operating theatre the scalpel blade slid
left to right along the length of the boy's sternum until a
trickle of bright red blood skidded over the plastic drape.
The electrocautery soon put a stop to that as it sizzled
down onto white bone, reminding me of that line from
Apocalypse Now – 'I love the smell of napalm in the
morning.' The whiff of white smoke told me that the
diathermy had too much power and I reminded the

orderly that we were operating on a child, not electing a pope, so would he please turn down the voltage.

Heart failure fluid was pushing up the diaphragm. I made a small hole in the boy's abdominal cavity and straw-coloured fluid poured out like piss into the wound. The noisy sucker removed almost a pint into the drainage bottle and his belly flattened out. A very quick way to lose weight. The saw zipped up the sternum, spraying beads of bone marrow onto the plastic. It breached the right chest cavity, releasing a knuckle of stiff, pink, waterlogged lung. Yet more fluid spilled out, so the sucker bottle had to be changed. It left no one in any doubt that this kid was seriously unwell.

Impatient to view the congenitally distorted heart, I dissected away the redundant thymus gland and sliced open the pericardium – the fibrous sac that encases the heart – with the same excitement and anticipation as unwrapping a surprise parcel at Christmas.

Everyone wanted to get a good look at the dextrocardia heart before I started, so I took a step back and relaxed for a minute. The plan was to open up the narrowed channel below the aortic valve by coring out as much solid tumour as possible, then close the hole in the atrial septum. I gave the order to go onto the heart–lung machine and proceeded to stop the empty heart with cardioplegia fluid. It lay cold, still and flaccid in the bottom of the pericardial sac. I gently pressed the muscle and could feel the rubbery tumour through the heart wall. By now I was sure that I couldn't reach it all with a conventional approach and that there was little point cutting into the ventricle that his circulation depended upon purely on an exploratory basis. So I told myself, 'Just do it.' Plan B. The eureka option, one that had probably never been done before. The perfusionist began to cool the whole body down from 37°C to 28°C. The boy was likely to be on the bypass machine for at least two hours.

At that point I had no option but to share Plan B with the rest of the team. I would chop out the boy's heart from his chest and, with it lying on a kidney dish full of ice to keep it cool, operate on it on the bench. Then I could twist and turn the thing as much as I needed to do a good job. I considered it to be a brilliant idea, but I had to work fast.

The process was equivalent to removing a donor heart for transplant then sewing it back into the same patient. Back in my research days I'd transplanted tiny rat hearts. This boy's heart should be no problem, even if the anatomy was unusual, so I transected the aorta just beyond the origin of the coronary arteries, then the main pulmonary artery. By pulling these vessels towards me, the roof of the left atrium was exposed at the back of his heart. I cut through the atria, leaving all the large veins from the body and lungs in place, then, lifting the ventricles out, I left most of the atria in situ. It was then, as you'd do with a donor heart, that I placed the cold, floppy muscle onto the ice.

Now I could see the tumour within the outflow part of the left ventricle. I started to dissect it out, cutting a channel through it so that it would no longer obstruct the heart. The tumour's rubbery texture was consistent with it being benign, making me optimistic that we had done the right thing. Both my assistants were shocked and mesmerised by the empty chest and were not assisting well. And the longer this heart remained without a blood supply, the more likely that it would fail when I re-implanted it. Frankly, the Australian scrub nurse was much sharper than these trainees, so I asked her to help. She knew instinctively what was required and injected the necessary pace into the procedure.

I was torn between just doing enough or making a radical job of it. But I wanted to tell the boy's mother that I'd succeeded in removing all of the tumour so I pursued it into the ventricular septum, close to the heart's electrical wiring system. I knew where this was situated in a normal heart, but its location was less certain in this case. After thirty minutes I infused another dose of cardioplegia solution directly into both coronary arteries to keep the heart really cold and flaccid, and fifteen minutes later the job was done.

I took the boy's heart back to his body, aligned the ventricles with the atrial cuffs and started to sew it in. I was really quite impressed with myself, the journal article already half-written in my head. The re-implantation process also closed the hole in the heart, so – with luck – he was cured.

This part of the operation had to be fail-safe as these stitch lines would be completely inaccessible in a beating heart. With both atria joined up again it was time to re-join the aorta and let blood back into the coronary arteries. The heart would start beating again and we could warm the boy's body up. All that was left to do was to reconnect the main pulmonary artery. By then the surgical assistants had also warmed up a bit, on familiar territory once more with the heart back where it belonged.

Usually a child's heart starts to beat spontaneously and quickly when its blood flow is restored, but this one was too slow. What's more, I could see that the atria and the ventricles were contracting at different rates. This told me that the conduction system between the two was not working, which is not good as a coordinated heart rhythm is much more efficient. The anaesthetist had already noticed this on the electrocardiogram but said nothing. After cooling, the conduction system often goes to sleep for a while then recovers spontaneously.

Ten minutes later and nothing had changed. I must have cut through the electrical bundle while dissecting out the tumour. Shit and derision. He'd need a pacemaker. This made me more anxious about another issue. A transplanted heart also loses its connection with nerves from the brain, nerves that automatically speed up or slow down the heart during exercise or changes in blood volume. This denervation, together with the disruption of the electrical conduction system, could be a real problem.

My earlier euphoria, optimism and self-congratulation quickly abated, and the young mother drifted back into my thoughts. This wasn't a good time to lose focus. There was still air within the heart chambers and it had to be let out, so I inserted a hollow needle into the aorta and

pulmonary artery. Air fizzed out of both. When air entered
the uppermost right coronary artery the right ventricle
distended and stopped pumping.

We needed another fifteen minutes on the bypass
machine for the effects to wear off. During that time I put
temporary pacing electrodes on the right atrium and
ventricle. We'd control his heart rate until the cardiologists could implant a permanent pacemaker. Gradually
the heart function improved. Obstruction gone, lungs
relieved of congestion, his life relieved of heart failure and
breathlessness. Or so I hoped.

The boy's pulse rate was only forty beats per minute,
less than half of what it should have been. We increased
that to ninety with the external pacemaker, and with this
improvement the blood started to well up from behind
the heart. I assumed that this was persistent bleeding
through my stitching, so I told the perfusionist to turn the
bypass machine off and empty the heart while I lifted it
up to inspect the join. Nothing. It looked great. No leak.

When we restarted the machine thirty seconds later
there was more blood. I inspected the joins of the aorta
and pulmonary artery. No leak there, either. Eventually
my first assistant spotted oozing from the aorta. The
needle used to evacuate air had gone through the back,
making a small hole. This would be inconsequential when
blood clotting was restored, so we separated the boy from
the heart–lung machine and closed the chest.

I didn't have long to reflect on our success as a message
came in from the adult cardiologists. They had just admitted a young male following a high-speed road-traffic acci-

dent. He'd not been wearing a seatbelt and his chest had impacted against the steering wheel with great force. He was in shock and his blood pressure could not be restored by fluid resuscitation.

Chest X-rays at the referring hospital had shown a fractured sternum and an enlarged heart shadow, and the veins in his neck were distended, suggesting blood under pressure in the pericardial sac. Not only that. The echo-cardiogram showed that the tricuspid valve, between his right atrium and ventricle, was leaking badly, hence the persistently low blood pressure and severe shock. The man needed urgent surgery, and could I please come and see him before it was too late?

I was distinctly uneasy about abandoning the boy but there was no choice. Leaving the operating theatre complex I found the mother sitting cross-legged in the corridor, alone and desolate. She'd been waiting there for five hours, and I sensed that she was about to implode mentally, her emotions bottled up for too long owing to her inability to communicate for whatever reason. And finally we'd taken away her bundle of rags. She saw me, sprang to her feet and panicked. Was the operation a success? I didn't need to speak. Our eyes met again, pupil to pupil, retina to retina. My smile was enough, and with it the message that her son was still alive.

Bugger protocol and the audience of cardiologists. I needed to show her some affection so I held out a sticky hand, wondering whether she'd take it or remain aloof. This act of kindness unlocked the tension. She grasped it and began to shake uncontrollably.

I pulled her in and held her tight, as if to say, 'You're safe now, we won't let anyone harm you any more.' When I let go, she held on tight and started to weep uncontrollably, waves of emotion discharging onto the hospital corridor and leaving my Saudi colleagues standing in an embarrassed silence. It took a while to calm her, and they were becoming increasingly anxious about their trauma patient.

I told her that her son would shortly leave the operating theatre, that they would bring him out in an intensive care cot attached to drips and drains, and that this might frighten her. She could certainly walk with them but not interfere. Again I sensed that she understood English, but in case she didn't one of the cardiologists repeated my words in Arabic. Then we left to review the injured man's echocardiograms, the ultrasound examination of his heart chambers.

By now the trauma patient was dying. He had a torn tricuspid valve, a rare, high-speed deceleration injury we never see with our mandatory seatbelt law. The right ventricle was pulverised as the sternum fractured and had been driven back against the spine, the rapid increase in pressure causing the valve to burst. Now, when the heart contracted, as much blood went backwards as forwards, little was passing through the lungs and the heart couldn't fill adequately because of blood in the pericardium. Cardiac tamponade, we call it.

Once I'd seen the pictures I didn't waste time visiting the patient. I just needed to crack that chest, relieve the tamponade and if possible repair the tricuspid valve. We had to get him onto the heart–lung machine quickly to

restore blood flow to the brain and correct his dire metabolic state. Then someone behind me whispered, 'Don't
rush. He's a madman. He killed the other driver.' I said
nothing. That wasn't my business. Striding purposefully
back to the operating theatre I encountered the little
entourage in transit to paediatric intensive care. The fast,
regular beeping of the heart rate monitor was reassuring.
Without diverting her gaze the mother held out her hand
as we crossed over, and I did the same. Contact.

I should have been with the boy in intensive care, at
least for the first couple of hours until I was confident that
he was stable. But now I couldn't be. Soon the trauma
patient was on the operating table being resuscitated. He
had disfiguring facial injuries and extensive bruising over
the chest wall, and the edges of the fractured sternum
were overlapped with a step deformity. But it was nothing
we couldn't fix with pins and wires.

Within minutes I had the chest open and was scooping
clumps of blood clot into a kidney dish. This improved his
blood pressure, but his right ventricle looked like tenderised steak – and it didn't contract any better than a steak
– and his right atrium looked like it would burst. So I put
the pipes directly into the major veins. As we started
cardiopulmonary bypass, his struggling heart emptied out
and flapped around at the bottom of the pericardial sac
like a wet fish. He was safe – and just in time!

With an incision directly into the right atrium the
ruptured valve was there in front of me. It was torn like a
curtain, but when I stitched it like torn cloth it was easily
repaired. I tested it by filling the right ventricle through a

bulb syringe. No leak. So I closed the atrium and removed the snares to fill it again. The job was done. The tenderised meat functioned better than anticipated and eased itself off the bypass machine. By then I'd had enough. I left my assistants to repair the fractured sternum and close the chest. No doubt he'd survive to go to prison.

The sun was setting on a hot and difficult day. For a while I felt content, satisfied after two 'out on the edge' operations, difficult cases that few heart surgeons would ever encounter in their whole career. I needed a beer, many beers, but there was no chance of that. I wondered whether the mother was happier now. She'd achieved what she set out for – treatment for her dying child.

Having heard nothing from intensive care I assumed that the boy was doing fine. Wrong. They were already in trouble. For some reason the doctors had tampered with the temporary pacing box and the electric stimulus from the generator had coincided with the heart's natural beat, fibrillating it and instantaneously inducing that uncoordinated, squirming rhythm, a herald of imminent death.

To counteract this they'd used external cardiac massage until a defibrillator was brought to his bedside. The vigorous chest compressions he'd been given had displaced the pacing wire from the atrium and, although the heart defibrillated at the first shock, the sequential pacing of atrium then ventricle no longer worked. Now only the ventricles could be paced. As a result there was a precipitous drop in cardiac output and his kidneys had stopped producing urine. The boy was deteriorating but no one had told me because I was in the middle of another big case. Shit.

Throughout this débâcle the poor mother had stayed by the cot where she'd watched them pounding on her little boy's chest, then witnessed the electrical shock that caused his little body to spring from the bed and convulse. At least he only needed one shot at defibrillation. The resulting beep, beep, beep was of little comfort to her, however, and like her child she was spiralling down.

I found her clasping his tiny hand, tears running down her cheeks. She'd been so happy as she escorted him from the operating theatre. Now she was desolate and I was too. It was clear that these intensive care doctors didn't understand cardiac transplant physiology.

And why should they? They'd never been involved with heart transplants so they failed to grasp that taking the heart out of the body cut off its normal nerve supply. They were pacing the heart at 100 beats per minute with an insufficient volume of blood while simultaneously flogging it with high doses of adrenaline to raise the blood pressure. This constricted the arteries to his muscles and organs, substituting blood pressure for flow and once again producing metabolic mayhem.

The nurse looking after the boy on the intensive care ward seemed anxious and was pleased to see me. A very capable New Zealander, she clearly did not rate the critical care registrar. Her opening remark was, 'He's not passing urine and they're not doing anything about it,' followed more directly with, 'If you're not careful they're going to fuck up your good work!'

I put my hand on the little boy's leg, the best way to judge cardiac output. His feet should have been warm, with bounding pulses. They were cold. He needed dilated arteries, less resistance to flow and less demand for oxygen. So I changed everything. Now the nurse was happier but the registrar was put out and phoned the on-call consultant. That was fine. I told the consultant to get himself in from home and discuss it.

We now walked the fine line between recovery and death. Much depended upon expert management, minute by minute, beat by beat, balancing the cocktail of power-ful drugs and maximising this buggered little heart's pumping capacity. The boy's lungs were inflamed and stiff after his long stint on the heart–lung machine so the oxygen levels in his blood were falling. Already the kidney failure warranted dialysis through a catheter inserted into the abdominal cavity using concentrated fluids to draw the poisons out through his own membranes. I needed the help of someone I could trust. Mayo man. I would stay in one of the on-call rooms a couple of minutes away where the residents slept.

The mother didn't want me to leave. She fixed her eyes on me, tears spilling over her high cheekbones. Profound separation anxiety was trying to pull me back but by then I was physically exhausted and fearful of how it would be if the boy died. She had no one else in the world, and although I wanted to be kind it was time to take a step backwards. Call it professionalism or self-protection. Perhaps both. So I reassured her that Mayo man was on his way, then I left.

By now it was well after midnight. The on-call rooms overlooked the rooftops, the club room opening onto a veranda under the night sky. Not as spectacular as the sand dunes at night, but good enough. There was juice, coffee, olives and dates. And Arab pastries. Best of all, a telescope for stargazing. I looked out at nothing in particular, wishing I could see England and home. And most of all my little family.

Now I tried to switch off. Mayo man knew I had more babies to operate on in the morning, so they'd call me only if it was absolutely necessary. I was desperate to find an improving child, with hot little legs and liquid gold in the urinary catheter. And I wanted to see his mother happy, cradling her boy in rags again.

I passed out in a heap, those deeply penetrating eyes still fixed on me, pleading with me to make things right.

Chanting from the minarets roused me at dawn. It was 5.30, and the fact that there had been no night call from intensive care was cause for guarded optimism. Today's operations were easy enough: holes in the heart to be closed with a cloth patch, careful stitching, then cured for life. Happy parents.

Soon I was thinking about the mother. How was she feeling now? I took tea out onto the roof as the blistering sun hauled itself into the sky, the air still cool and fresh, the temperature bearable.

At 6 o'clock Mayo man called. After a pause with heavy breathing he said, 'Sorry to wake you with bad news. The boy died just after 3 am. Quite suddenly. We couldn't get him back.' Then silence in anticipation of my questions.

I'd had calls like this my whole career but this one made me miserable. I asked what had happened. At first the boy had started fitting, perhaps in response to the metabolic mess and high temperature, quite violent fits that were difficult to control with barbiturate drugs. The acid and potassium in his blood were still high because dialysis had not yet been started. And then when he had a cardiac arrest they couldn't get him back. Mayo man had been hesitant to wake me with bad news and he was sorry for my loss.

A kind thought, but what about the girl? Did they want me to come across and try to communicate with her? Mayo man didn't think that would do any good. Once again she'd been beside the cot during the resuscitation efforts. She was now obviously very distressed, and had been hysterical beyond reason when told that her child had died. They'd moved the cot into a single room away from the unit, where she could hold him and grieve in private. All the catheters, drains and pacing wires had to be left in place until the autopsy. I felt bad about that. How could she cuddle the lifeless infant with plastic tubes emerging from every orifice?

This is cardiac surgery. Another day at the office for me, the end of the world for her.

I was drawn to her like a magnet but had to stay away. In an hour I was due back in the operating theatre, needing to be on top form for someone else's baby. Another mother who cared just as much. What a fucking job. I was a sleep-deprived, psychological wreck operating on tiny babies on the other side of the world.

I called adult intensive care to enquire about the trauma patient, the man who'd recklessly crashed his car and killed another driver. He was doing fine. They were going to try to wake him up and take him off the ventilator. There was a certain irony in that. Thinking of the boy, I just wished it had been the other way round. Forbidden thoughts. Surgeons are meant to be objective, not human.

I took my despair to the canteen, where I glimpsed the miserable paediatric registrar devouring breakfast. My instinct was to avoid him, but it wasn't his fault. It was me who had done the surgery and I regretted not staying up all night to see it through. When he saw me I could tell he was bursting to tell me something.

He told me that the mother had disappeared from the room, taking her dead child with her. No one had seen or heard her take off, and there had been no sight of her since. I uttered one word – 'Shit.' I didn't want to continue that conversation. I assumed she had taken off into the night like her flight from Yemen, but this time carrying a lifeless bundle. By now she could be anywhere and I was anxious for her.

I heard the news as I was stitching the patch into the first ventricular septal defect. When the Saudi hospital staff turned up for work they'd found them, two bodies lying lifeless in a heap of rags at the bottom of the tower block. She'd removed the drips and drains from the little body before leaping into oblivion to catch up with him in heaven. Now they were both together in the cool of the mortuary, inseparable in death. Two hundred per cent mortality for me.

Most writers would end this tragic tale with the mother's suicide and the discovery of the bodies at the foot of the tower block. A devastating end to two fragile lives. But real-life heart surgery is not a soap opera. The job goes on, and there were too many unanswered questions. I always attend the autopsies of patients I operate on. First, to protect my own interest – to make sure the pathologist understands what has been done and why – second, as a learning experience to see if anything could have been done better.

To spend all day, every day with the dead makes mortuary people different, as I knew from my time at Scunthorpe War Memorial Hospital. The technicians work like butchers, slicing open the carcass, removing the entrails, sawing off the cranium to lift out the brain. Here an ageing Scottish pathologist ruled the roost. Resplendent in green plastic apron and white Wellington boots, sleeves rolled up, cigarette dangling from the corner of his mouth, he was grunting away to himself, documenting the cause of death of the man killed by my trauma patient. Fractured neck and brain haemorrhage, together with ruptured aorta – high-speed crash injuries. I was new to him as surgeons didn't visit the mortuary very often. Mercenaries were rarely interested in learning from their failures.

That morning there were seven naked corpses lined up on separate marble slabs. My attention was immediately drawn to the mother and child on tables side by side, so far untouched. I explained to the Scotsman that I was pressed for time. He was grumpy but cooperative as the technician joined him. Only the child was officially my

patient. His head had hit the ground first and the skull
was split open, with the brain shattered like a fruit jelly
dropped onto the floor. There was little blood because he
was already dead. I had an important question about the
brain. Did the boy have tuberous sclerosis, the brain
pathology that goes with rhabdomyomas in the heart?
This causes fits and could have precipitated his death.

I re-opened the chest incision myself, unpicking the
stitches. Was I correct about the disconnected pacing
wire? It was difficult to know because the mother had
pulled it out after he died. But there was a clue: a blood
clot squelched out from beside the right atrium. In every
other respect the surgery had been successful, the tumour
had virtually gone and the obstruction relieved. The
Scotsman dropped the heart into a jar of formaldehyde
and kept it on the shelf as a rare specimen.

Eager to maintain the momentum, the technician sliced
open the abdomen and eviscerated the child. All organs
back to front, floating in heart-failure fluid but otherwise
normal. Cause of death: congenital heart diseases – operated. A second technician came along, stuffed the brain
and guts back into the abdominal cavity, then sewed the
boy up. After repairing the rent in his head, the boy was
disposed of in a black plastic bag. End of story. Blood and
body fluids were washed from the marble slab, and then
no further trace of his tragic short life remained. There
was no one to bury him.

I was drawn to the mother's jet black broken body,
now naked on the adjacent slab. So thin. Still proud.
Mercifully her beautiful head and long neck were intact,

her once-sparkling eyes wide open but dimmed and fixed on the ceiling. Her injuries were obvious without slicing her open: arms broken, legs horribly distorted, abdomen swollen from liver trauma. No one survives such a fall and she knew it. How different all this could have been had the boy survived. The happiness she'd have experienced watching him grow up with a heart that worked. I watched the technician fold the scalp over her face and remove the top of the skull with a circular saw, lifting the lid on her tragic memories. Why did she never speak?

Like an archaeological dig, the vital clues emerged. Above the left ear she had a healed skull fracture with damage to the dural membranes and brain beneath. This involved Broca's area, the zone of the cerebral cortex responsible for speech. When the Scotsman carved her soft brain into slices the scar became more apparent, extending deeply and slicing through the nerves to her tongue. These were injuries she was lucky to survive during her abduction in Somalia and the reason she never spoke – why she understood but couldn't respond.

I'd seen enough. I didn't want to watch her eviscerated, her life blood spilt out onto the mortuary slab, or witness her ruptured liver and fractured spine. She died from internal bleeding but I remember thinking that a fatal head injury would have been kinder for her and that she'd have been better off dying in Somalia, sparing her the misery of her life in South Yemen. With that I thanked the Scotsman for his cooperation and headed back to the operating theatre where I belonged, hoping for a better day, desperate to do some good.

6

the man with two hearts

A successful cardiac surgeon is a man who,
when asked to identify the three best
surgeons in the world, has difficulty in
naming the other two.

Denton Cooley

IT WAS QUITE BY CHANCE that I ever met Robert Jarvik. I'd travelled to San Antonio in Texas for the 1995 meeting of the US Society of Thoracic Surgeons and it was there, while wandering around the Alamo, that an executive in the cardiovascular industry asked me to give an opinion on a new product. He took me back to a corporate meeting with an engineer whose name I was well familiar with – Robert Jarvik.

The device under consideration was a small turbine pump designed to boost blood flow down the legs of patients with severe peripheral arterial disease. When the company men moved off to their dinner with clients, Jarvik turned to me and said, 'Come up to my hotel room.

I have something interesting to show you.' I'm always wary of such invitations from men, but on this occasion I was intrigued.

First he filled the sink in the bathroom, then he took out a small plastic container from his briefcase. It looked like a sandwich box, and inside was a thumb-size titanium cylinder with an attached vascular tube graft and a silicone-covered electric power cable. He put the titanium cylinder in the water, attached the cable to a telephone-sized controller and switched it on. Whoosh!

This small, continuous-flow pump shifted around five litres of water a minute, redirecting it through the graft back into the sink without noise or vibration. Jarvik had been working on the concept of a left ventricular booster pump for many years, one designed to be 'functional but forgettable' for the patient.

I then said something stupid. 'This is a great pump for water but if you put it into the bloodstream it will clot or chew up the red cells,' as if Jarvik had neither thought about these issues nor addressed them. Then I said something sensible. 'But I'd love to work with you to test it, away from the Food and Drugs Administration [FDA]. If it looks good we could use it in the UK long before you'll get permission over here.'

This was a punt in the dark, so I continued by asking if he was already working with a centre in the US. He said that he was testing the product with Bud Frazier, the head of the transplant service at the Texas Heart Institute. Bud was America's foremost advocate of mechanical circulatory support devices. Jarvik told me that he was at the meeting and asked whether I'd like to be introduced. So off we went to find him.

Bud was 100 per cent Texan, wearing a Stetson and cowboy boots with a smart suit. Both charming and understated, as well as being a surgeon he was a collector of antiquarian books. He expressed confidence in the new pump, which was currently known as the Jarvik 2000, the year 2000 being the projected date for human implants if the laboratory research work went well. He asked if I'd like to see the calves implanted with the pump at the Texas Heart Institute. The Texas Heart animal laboratories were considerably more impressive than my facilities for humans and were filled with sophisticated, modern equipment that I could never get for my patients.

When I visited I found the calves happily munching away at hay in the stalls. The monitors showed the impeller to be spinning at 10,000 rpm, pumping around six litres of blood per minute, more than what was needed by a patient at rest. Bud handed me a stethoscope so that I could listen to the faint, continuous whine of the turbine in the bloodstream.

I'd been wrong. It didn't damage the blood cells and, irrespective of the lack of blood-thinning therapy, it didn't clot either. This was a complete revelation. Could the device be a monumental step forward for patients dying from heart failure? It was certainly right in front of my nose with an opportunity to get involved. I took the chance and offered to test the Jarvik 2000 in sheep back in Oxford.

As a result of these fortuitous meetings I returned to Oxford thinking that an immense international project would soon be underway: Houston, Jarvik Heart from New York ... and Oxford. Indeed I felt as though I could fly back to London without the aeroplane. But after I'd reflected on it a little longer I was no longer so positive. After all, I had no research funds and no access to a large animal laboratory. All I possessed was grim determination and the will to succeed.

Within months I had raised sufficient funds from philanthropists to begin the project. Now Cambridge had their pig-heart transplant programme and Oxford had miniature artificial hearts – a veritable Varsity match. Soon we confirmed what Houston had suspected: continuous blood flow without pulse pressure was safe and effective. This changed the whole philosophy of blood-pump design, eliminating the need to replicate the pulsatile function of the normal human heart.

Against the background of this thriving research programme I felt justified in beginning a surgical heart-failure service in Oxford. There were many thousands of terminally ill heart-failure patients in Britain, but fewer than two hundred heart transplants each year. Most patients with deteriorating kidney and liver function were considered too sick to be accepted onto a waiting list. Their lives would be terminated by drugs in the name of palliative care. My vision was that these desperately symptomatic patients should be helped by 'lifetime' blood-pump support – an 'off the shelf' mechanical solution that didn't need a dead person or frantic donor-heart retrieval by helicopter in the middle of the night. My megalomania told me to establish Oxford as a national centre for mechanical-circulatory support.

In Houston, Bud was already implanting a more conventional pulsatile ventricular-assist device to keep patients alive until a donor heart could be found. A so-called 'bridge to transplant', the ThermoCardio-systems HeartMate pump was intended to replace the diseased left ventricle by filling and ejecting blood rhythmically. It was shaped like a round chocolate box and too large to fit in the chest, so it was implanted into a pocket in the abdominal wall. Out of this emerged a stiff electrical cable to the external batteries and controller. This 'life line' also incorporated an air vent that hissed continuously in time with the pumping mechanism and was audible from across the street.

The prolonged hospital confinement – as the average wait for a donor heart for patients with the HeartMate assist device was 245 days, and much longer for those with blood group O – was massively expensive and psychologically damaging. But with increased experience, the Houston team gained confidence that the patients should be released from hospital. Not only that – they considered that this mechanical blood pump could be used as an alternative to heart transplantation.

Bud knew that the FDA would not consider it as a permanent treatment at the time. He phoned me in Oxford and said that, since we were working together on the Jarvik 2000, might we be able to test the 'Lifetime Support' concept in NHS patients with the HeartMate.

ThermoCardiosystems would provide the pumps free of charge, and this would offer a lifeline for terminally ill patients who'd already been turned down by the transplant centres, who were breathless on minimal exertion, swollen with fluid and housebound. In other words the walking dead, though rarely walking!

This was the opportunity I'd been waiting for. I flew to Houston to watch an implant and meet the transplant candidates who were living with the device in the hospital. When asked whether I'd like to assist with an operation I jumped at the chance. The patient was a Mid-Western college football player who'd caught a virus – virile to viral, athlete to asthenic. The poor lad was wasted and waterlogged, his life ebbing away inexplicably. His girlfriend sat by the bedside beside him and it seemed that she didn't know what to say. What *do* you say to someone who needs an artificial heart?

She was a football cheerleader, very pretty, but with nothing to cheer about now as her hero was dying. She'd watched him deteriorate, lose his place in the team and drop out of college. But it had taken too long to realise that he was sick and that it wasn't recreational drugs, as some had suspected. And what should she do now? Leave him and get back to her studies, or stick by the lad whose best chance would be a heart transplant? Life is a bugger sometimes and we seldom stop to think how it is on the other side. For good reason, I guess, as it rarely helps.

In the operating room the nurses helped the surgeon to gown and glove, then painted and draped the patient, leaving the entire chest and upper abdomen exposed.

Once he rippled with biceps, pecs, abs – the lot. Now he was just skin and bone, with distended liver bulging beneath his ribs. Heart failure is shit. The creeps who declined to fund our research should come and stand here at the operating table.

Bud made the incision from the lad's neck down into his belly as the HeartMate pump needed a sizeable pocket in the abdominal wall, making it look like an alarm clock beneath the skin when in place. The dilated heart was huge and the left ventricle barely moved, the usual yellow fluid spilling out of the pericardial sac and filling the new pump pocket until it disappeared down the sucker.

While I was despairing at the utter decline of this great athlete, Bud was concentrating on where to bring the stiff electric driveline through the skin, looking for a position that didn't interfere with his belt and trousers, and where he could keep it clean, with as little motion as possible. He made a stab wound with the scalpel and we hauled the line through. More than a centimetre thick – and stiff enough to prevent kinking of the air vent – this wasn't a domestic electric light cable. It was his lifeline, as vital as the placenta to an unborn child. Then we carefully stitched the pump outflow graft to the ascending aorta as it left the heart, ensuring it was precise, otherwise it would bleed profusely under pressure.

All that remained was to sew a restraining cuff to the apex of the heart and use a circular coring knife to make a dollar-sized hole for the pump inflow cannula. Now blood returning to the heart from the lungs would pass straight through the mitral valve and into the machine, his own buggered ventricle completely redundant. But I was thinking about Jarvik's new device, which was barely larger than this pump's inflow cannula. The titanium shell of the HeartMate's pulsatile pumping chamber was huge in comparison.

Before switching on the HeartMate pump it must be filled with blood to expel air. 'Air in the brain, life down the drain,' I quipped. Poetic, perhaps, but by then I was jet lagged, sleep deprived and a touch manic. The technical team had made the connections, we were ready for the big switch on, and, as the pusher plate mechanism started to shift in the pump housing, air hissed to and fro in the vent like a steam train setting off. The chamber filled, then ejected blood into the aorta, any residual air fizzing and frothing through needle holes in the graft, gratefully making its exit. His own useless muscle was sucked down, no longer tense and quivering in an attempt to keep him alive. He had a new heart. Just temporary but I hoped it would serve him well.

I wondered how his girlfriend would react to this pulsating, hissing monster inside him and to the stiff new appendage emerging from his belly. How long would she stay with him now? These were thoughts that I'd never normally entertain, this lack of empathy stemming from continuous stress and fatigue. If I ever saw her again I resolved to be really supportive and tell her how well it had gone. That he would now get better and stronger. Soon someone would have their brains blown out in Houston. Then – if he was lucky – he could have their heart.

It took a while to stop the bleeding and the general oozing from the poorly functioning liver and bone marrow in common heart failure patients. Bleeding, excessive blood transfusion, then bad lungs and kidneys were a common scenario in such surgery. Now I needed to go to the airport for another twelve-hour flight back to a completely different world where none of this would have happened, where he'd have been left to die. But I wanted to see his girlfriend first. The young man's parents had just joined her, all of them anxious as hell together.

As she looked up and recognised me I told her quickly that the operation had gone well, something that always triggers waves of relief, five words that cut straight through the tension. The coiled spring was released, her sweet face lit up with joy and she started to cry. So she really did care about him, not just that he was a football star. I felt like a miserable shit for doubting that. His parents hugged and thanked me. 'For what?' I thought. I just assisted Bud. But with good news, gratitude goes out to all. I wished them all the best and a donor heart soon. With all the misery that would bring for another family.

With the help of Professor Philip Poole-Wilson at the Royal Brompton Hospital we soon identified potential candidates for the HeartMate pump in London. Sadly, the first and youngest died before we could help. The next, however, seemed ideal. He was sixty-four, tall and slim, and had already been turned down for a transplant. Like the American football player, he had dilated cardiomyopathy, possibly genetically linked, but more likely the result of a virus or an autoimmune disease. An intelligent Jewish man, Abel Goodman had a huge heart and was virtually bedridden.

On the plus side, his coronary arteries were free from disease and he still had reasonable kidney and liver function. Hopefully that would make the post-operative care less of a battle – and less expensive. His breathlessness was worsening, so he was propped up in the bed with pillows as he was unable to lie flat with his swollen legs and abdomen. Philip needed to admit him to the Brompton for medical stabilisation so I went to see him there. I always loved returning to that hospital, this time as my own man, more proper heart surgeon than music-hall joke.

Abel sat bolt upright in bed, with laboured breathing, sweat beading on his brow and the fear in his eyes that says, 'I'm not long for this earth.' He was too distressed to talk. Too sick for a haircut, as we say. Ready to meet his maker but secretly hoping his saviour had come instead. I shook his limp hand. It was cold and slippery, his blood not reaching that far. I explained that the HeartMate pump I'd seen in action in Houston would take away his terrible symptoms and that he was the first patient in the world to be offered this technology on a 'lifetime' basis. Normally it was only used for transplant candidates. How long was a 'lifetime'? I didn't know, but without it he'd probably die within weeks. At most. (I thought he could 'go' during the conversation.)

His head tipped back and his eyes rolled as he digested the information. There wasn't much blood reaching his brain either, but he managed to raise his head from the pillow and murmur, 'Let's get on with it, then.' I think he hoped it would be that day. Enough was enough.

It was 3 pm in London, six hours ahead of Houston. I called Bud to explain the tight timeline and that we'd only be given permission to use the pump in a dying patient on 'humanitarian grounds'. We had the dying patient, so could we do it next week? The line went quiet for what seemed like minutes. Then one word: 'Yep.'

I felt a surge of adrenaline and excitement. We'd implant a mechanical heart in Oxford. But who was I excited for: Abel or myself? I was an ambitious bastard and we all wanted to do something special, to take risks – not just for the patients but for ourselves, knowing that it would create headlines, as well as extreme animosity from the transplant lobby, flying in the face of that curious attitude that it's better to let patients die than attempt something new.

The Houston team arrived in Oxford on 22 October. That evening the anaesthetists, perfusionists and nursing teams gathered in the conference room. We needed to talk through the procedure and get acquainted with the equipment, not to mention my Texas friends and their dress code – cowboy boots and Oxford colleges rarely come together.

Abel had survived the transfer out of London and was now bemused by the cosmopolitan medical team, yet too breathless to care. The nurses told him to think positively and the ward orderly took his order for supper the next evening. He didn't want the gammon. The rabbi came to prepare him for death.

Bud had never been to Oxford. With his interest in antiquarian books I wanted to show him the Bodleian library and the ancient colleges in the centre of town, as if on a different planet from Houston. We drank beer in the Eagle and Child tavern where Tolkien and C. S. Lewis met regularly on Thursday evenings in the 1930s. I listened to his stories about the Vietnam War, how he'd been a helicopter medic in the middle of the action and would sit on his helmet to avoid getting his testicles blown off. Several of his surgical colleagues didn't make it. Bud kept his balls – and it showed. He'd done more heart transplants than anyone else and more ventricular assist devices. He reminisced about the agony and the ecstasy of those days, all when I was still at medical school.

Then I asked how the college footballer was doing. He was still wandering the corridors of the Texas Heart Institute, no longer in heart failure and building up muscle again. But still there was no donor. His girlfriend had gone back to college.

For me this evening was the calm before the storm. And Bud hoped it would be the beginning of a new era, one in which pumps would be used to treat patients who had no other option. Why should these medical life-savers be inextricably linked to transplantation? It was a waste of life-saving technology, thousands of dollars thrown away when the transplant happened. I wondered what other historic discussions had taken place in the Eagle and Child over the centuries. This must have been the first about artificial hearts.

When the following morning came it was all a lot more relaxed than I'd expected. The pump company representatives sat chatting with Bud in the operating theatre coffee room. His technical assistant Tim Myers was already laying out the equipment with the nurses, who were excited but nervous, not wanting to screw up in front of their distinguished visitors. Abel came down from the ward with a procession of family and friends to see him off. Off to where, was the question. He sat slumped forward on the trolley in a white gown, head bowed, hands on his skinny knees, gasping for breath in anxiety. He just wanted to be put to sleep. When they passed me in the corridor he raised his head tentatively and murmured, 'See you later.' The man remained optimistic to the last.

This time I'd do the surgery, with Bud assisting and my colleague David Taggart joining us as second assistant. For such a politically charged event we managed to stay calm and business-like, almost to the point of levity. The pump's manufacturer had realised that surgeons were not the brightest members of the medical profession so they'd put arrows on the titanium pump housing to make sure we implanted it in the correct orientation. I enjoyed making the gigantic incision from neck to umbilicus, never having been one for keyhole surgery, but while I was proud of my own abilities I was embarrassed by our outdated equipment. The old saw juddered up the breast-bone, almost failing to reach the top. We made the pump pocket in the upper left abdominal wall, then zipped open the tense pericardium to expose Abel's huge heart.

Like a baptism, I worked through the implant process step by step, doing it Bud's way. Pipes in for cardiopulmonary bypass, next onto the heart–lung machine, empty Abel's heart, then carefully sew the restraining cuff to the apex of the left ventricle and the vascular graft to the aorta. We cored out the disc of sick muscle from within the cuff and kept it for the microscope. Then in went the inflow cannula of the pump. Job done.

The critical last step was to remove all air from the system before the big switch on. We filled the heart by cutting back on the flow from the heart–lung machine. The left ventricle filled and blood entered the pump through the inflow. Air was pushed on into the vascular graft and expelled through a wide-bore needle. With the titanium 'chocolate box' sitting securely in its pocket, Tim was told to 'switch on'. The noisy mechanism started up with its characteristic hissing noise and the last few bubbles fizzed out of the air needle. Abel had a powerful new left ventricle, one that you could hear across the street – but the patients get used to that, just like mechanical heart-valve patients get used to the tick, tick, ticking in the dead of night. It becomes part of bionic life and is much preferable to the alternative. Usually.

Abel woke quickly from the anaesthetic. Perhaps too quickly. He was immediately taken off the ventilator and his tracheal tube removed. I could tell that he felt different. He had a twinkle in his eye and a cheeky grin, and showed the relief and bewilderment that everyone has when they wake from an anaesthetic – the 'I'm alive' moment. All of his four limbs moved normally and he had no neurological issues. I felt like calling the chief executive – just as Christiaan Barnard had done after his landmark transplant – to say, 'Sir, we've implanted an artificial heart and the patient is fine.' But something told me to hold back and remain cautious for a change. This wasn't about me, it was about getting Abel back on his feet again and I was worried that his blood pressure was too high. Rather than his own feeble left ventricle, he now had a powerful machine driving the circulation and was pouring out his own adrenaline in response to the unknown. The intensive care doctors needed to give him vasodilator drugs – and anticoagulation for his own heart's abnormal rhythm – then sedate him for the night. The aftercare was as important as the surgery. I needed some sedation too, but overall it had been a great day.

No news is good news and I heard nothing overnight. Always on a tight schedule, Bud and the company men left for Heathrow early the following morning and I drove to the hospital at 7 am, filled with optimism and self-congratulation. I composed a press statement in my mind, fantasising about the headlines: 'Oxford surgeon implants artificial heart' or 'Dying man saved by heroic surgery'. So I deserved the shit that hit me when I reached the bedside. I could see it in Abel's face – that vacant look. He was drooling from the right side of his mouth, his eyelid drooped, and he didn't greet me with enthusiasm and gratitude as I'd hoped. He couldn't even lift his right arm or leg. He'd had a fucking stroke.

Every possible expletive ploughed through my cerebral cortex while the pump hissed at me. He was pink and warm, with great blood flow – but fucking paralysed. After everything had gone so well. Why had no one warned me? I instinctively wanted to blame someone else. But for what? My gut feeling was that he'd thrown off a blood clot – either from his own heart or from the foreign surfaces of the pump or vascular graft – in which case we should give him the rapid-acting anticoagulant heparin since the warfarin wouldn't have had time to take effect. But a neurologist colleague persuaded me to do a head scan first, to document the extent of the brain damage and rule out a cerebral haemorrhage. If we gave heparin after a bleed into the brain it would surely prove fatal. But whatever the cause, this was a catastrophe, not least because of the financial implications of prolonged intensive care, all to be paid for by my research funds.

I accompanied Abel to the scanner. Bud and his team were already at Heathrow, unaware of the miserable development, and I was too pissed off to call them. Let them enjoy the flight back. I watched the scanner construct slices through the brain. The pathology was obvious but unexpected. There *was* bleeding into the brain. Not only that. The bleeding originated from an area of previous stroke that was definitely not recent, perhaps many months old. Why didn't we know about that? It transpired that Abel's wife had no knowledge of it either. He'd had headaches from time to time but had never suffered paralysis or weakness. Before now. So it must have been a 'silent' stroke, leaving us with the 'devil versus the deep blue sea' dilemma. Damned if you do, damned if you don't. For now, Abel was disabled, but he wasn't going to die. It was either 'Think positive', or get out of the high-risk business altogether.

I flicked the switch. Abel needed cardiac- and neuro-rehabilitation. Many strokes recovered with time and effort. He couldn't swallow, so we needed to feed him through a gastrostomy tube. The gastroenterologists inserted this directly into his stomach through the abdominal wall. He couldn't cough adequately, so he needed frequent chest physiotherapy. When he developed pneumonia, he had antibiotics. When he coughed so hard that he disrupted the skin around the drive line exit site, we revised it surgically. The physiotherapists worked hard to mobilise him. In three months the paralysis abated to weakness and the weakness resolved with exercise. Soon he was on the move again, rehabilitating himself. His speech returned, his swallowing improved and he restlessly wandered the hospital corridors, no longer breathless or swollen with fluid, no longer in heart failure. His life was returning, and so was my determination to press on.

From the sound of the pump and the hissing of the air vent – like a snake, but sixty times a minute – we always knew that Abel was close by before we saw him. It wasn't easy for him to live with, but much, much better than breathlessness. One day I passed him while he was sitting out in a chair. He volunteered that he felt below par. When we persuaded him to get back into bed and attached him to the monitor we saw the reason for this. His own heart was in ventricular fibrillation, that uncontrolled rhythm that's immediately fatal in an unsupported patient. Despite the fact that his right ventricle was now

functionless, the left ventricular assist device kept him
going. Incredible, I thought. Yet it happened on five
separate occasions and we just defibrillated him each
time. A quick sedative, apply the paddles and zap! – his
own heart started again. In time we noticed another thing.
His heart was shrinking and contracting more vigorously,
replicating Bud's finding that the dilated cardiomyopathy
heart gets better with rest. It was important to find out
why this happened on a molecular basis.

Had Abel died from stroke our charitable support
could have died with him. As it was, he survived and was
rehabilitated. The HeartMate continued to work well and
we were close to discharging him from the hospital. Then
the next patient was referred.

His name was Ralph Lawrence, and he'd taken early
retirement from his job as a finance audit manager with
Rover. He and his wife Jean liked to dance – folk dancing,
barn dancing, ballroom dancing – energetic stuff, and
they liked to travel around the country in their caravan.

Then in his early sixties, Ralph found himself increasingly breathless. The chest X-ray showed an enlarged
heart, so his local hospital in Warwickshire referred him
on to the heart failure clinic at the Royal Brompton
Hospital, where Professor Poole-Wilson diagnosed dilated
cardiomyopathy. The first step was treatment with heart
failure drugs, followed by what was a new treatment in
those days – electrical cardiac resynchronisation therapy
with a special pacemaker. The aim was to better coordinate the contraction of different parts of his dilated heart

to make the whole organ more efficient. But the beneficial effects can wear off as the heart gets bigger, and he was now in trouble again, severely symptomatic with a poor prognosis. Could he have a transplant? When Ralph was told that there was no chance of one at his age, strangely enough he accepted it, agreeing that scarce organs should go to younger people. He was a very likeable man with a supportive family, and we thought him an ideal candidate for the HeartMate.

While he was unable to do anything, Ralph was stable and not as sick as Abel Goodman had been. He had a few weeks to consider and reconsider the prospect, and we gave his family the HeartMate patient guidelines to look at. These were a daunting read, even for those who could anticipate a transplant in time. No swimming or baths. Showers, fine, as long as the electrical equipment was covered. Avoid tight clothing or dressings that might bend or kink the vent tube. Always have the emergency back-up equipment available. If the yellow spanner lights up on the controller it signifies malfunction. A red heart symbol with audio alert means loss of pump support, seek immediate assistance. And so on. This was all worrying stuff that Abel had had no time to consider.

I saw Ralph with Jean in my office in Oxford. They weren't easily put off by the literature as by now his life was intolerable. There were no more outings, and he was sleeping propped up in a chair, ankles and feet too swollen for shoes, likely to die suddenly at any time. And the family knew that. I was concerned that he was an insulin-dependent diabetic, but he managed that well, being used to taking responsibility for his own health. He had a positive attitude and wished to proceed as quickly as possible.

'So why not start today?' I said. I thought they should arrange to meet Abel to ask him how he felt about life with the 'alien' inside him. I knew what his answer would be: 'Better than heart failure. Better than being dead.' And Jean needed as much knowledge of the HeartMate as her husband, as she might have to cope at home in an emergency, perhaps even work it manually for him during a power cut.

We agreed on a date for surgery, a Wednesday just four weeks away, giving us time to make arrangements with Houston. But on this occasion there was one more consideration. The grapevine had spread extensive awareness of Abel's operation. Given his stroke we'd tried to stay low-profile, but because we were planning Ralph's operation a month ahead it was inevitable that information would be leaked to the press. This was a two-edged sword. Public awareness helped me to raise the charitable money needed to maintain the programme but bad publicity in the event of the patient's death could kill us. These debilitated heart failure patients would never be offered a hernia operation, yet alone heart surgery. So how to control the risk?

Between us we agreed that one newspaper should be given access to Ralph's operation to avoid a media scrum. First and foremost, the family needed peace when – or if – he should leave hospital. The *Sunday Times* was the chosen route. They could have in-depth access to the whole undertaking, as long as the patient and family were treated discreetly. In return we'd be grateful if they'd consider a charitable donation. Not payment. But without charity, Ralph wouldn't get his operation.

The night before the operation Ralph and Jean stayed together in a room provided by the hospital. Jean told the *Sunday Times*, 'We were quite rested. He had come to terms with everything and was just happy that the operation was to go ahead.' At 9.30 am on the Wednesday morning a sedated Ralph was taken by trolley to Theatre 5, once more unable to lie flat without gasping for breath. We hoped that he'd never again have that grim feeling of choking to death. This time there was huge interest within the hospital, so we agreed to video the operation and relay it to an auditorium. I was happy for the journalists and hospital managers to watch. In surgery we have a saying, 'See one, do one, teach one.' I'd seen one in Houston, done one in Oxford, but was quite sure that I was not about to let anyone else do Ralph's operation under instruction. Bud and I waited quietly in the coffee room while the anaesthetist put Ralph to sleep.

It was 5 o'clock in the small, stuffy waiting room. All day, every day it was 5 o'clock because the clock had stopped long ago. Only the growing stack of empty polystyrene cups marked the slow passing of time. Jean sat waiting for news, poleaxed by hand-wringing anxiety. Then at 2 pm came the news that she'd been waiting for. Ralph was being wheeled back to intensive care.

On 12 May 1996 an X-ray of Ralph's chest and artificial heart filled the whole front page of the *Sunday Times Magazine*. The caption read: 'The man with two hearts –

why a lump of titanium, polyester and plastic is ticking away inside Ralph Lawrence'. It was a risk to give a front-line national newspaper direct access to an artificial heart operation, with pictures from the operating theatre, and interviews with the family and staff. But they presented it extremely well and everyone could read it – the prime minister, members of parliament, even the Queen. The paper produced a pictorial, blow-by-blow account of the operation, which helped us sustain our laboratory research programme. Although we'd struck a chord with those who saw innovation as the duty of the NHS, we'd not done so with the NHS itself. This technology cost money and there'd be no support for it.

We always felt that it was Abel's high blood pressure that caused the cerebral haemorrhage, so we kept Ralph profoundly unconscious for several hours. It was the middle of the night when he regained consciousness and Jean was sitting there by his bed, watching the visible action of the pump thudding away in his belly amid the paraphernalia of intensive care. Through the oxygen mask he said something to Jean. 'You're thirsty?' she queried. 'No. Is it Thursday?' came the reply. Two days later Ralph was out of bed, sitting in a chair. The next day, Saturday, he was walking around the intensive care unit with the physiotherapist whose job it was to rehabilitate him.

Then disaster. As I was jogging through Blenheim Park my mobile rang. Abel was in great pain and suffering haemorrhagic shock on the ward, with acute bleeding around the pump. This had caused a huge swelling beneath his ribs, just at a time when his own heart had virtually recovered. We needed to remove the pump straight away and stop the bleeding, or he was going to die. I told them to call in the theatre team immediately.

I ran home faster than I should have done at my age and jumped into the car. The roads were quieter at the weekend but I was pessimistic whether we'd get him opened up in time. Either we would or we wouldn't – one had to remain sanguine, and an agitated, overexcited or anxious surgeon would never succeed in this predicament. I worked out what to do in the car. We'd never be able to reopen his chest quickly without causing damage, so I'd have to expose the artery and vein in his groin, cannulate both and begin cardiopulmonary bypass. Then he'd be safe. With enough transfused blood we could maintain flow to his brain and switch off the HeartMate. We managed to do this just in time, his blood pressure having fallen to half normal despite the transfusion.

I pulled the wires out of his sternum and ran the oscillating saw up the middle of the bone. As the edges parted, strips of shiny purple blood clot slithered through the gap and bright red blood spilled out from the lower end. I soon figured out that the change in Abel's heart size had probably altered the position of the HeartMate inflow cannula and that this had sheared open the apex of the now smaller heart. My informed guess had turned out to be correct. As I dissected open the inflammatory mass I could see that the join between the vascular graft and the aorta was secure.

It was a straightforward decision. The pump had to come out. Either Abel's own heart would succeed in supporting his circulation or he was dead. The easiest way to stop the inaccessible bleeding was to cool his body to 20°C, then stop the circulation altogether. In the meantime I amputated the HeartMate power line and discarded it, scooping out a mass of blood clot from the pump pocket in the abdominal wall. We were making progress, but I reflected that it wasn't a great way to spend the weekend.

At first it came as a bitter blow to his family. They were looking forward to having him home after five months in hospital while he was in good form. Abel and Ralph's wives were waiting together, one hoping for a miracle, the other now realising that a successful implant didn't mean happy ever after. Bad news travels fast and a sombre mood soon spread through the hospital. Abel's nurses and physiotherapists thought they'd lost him after months of intense effort to get him over the brain haemorrhage, which would have been a tragedy for us all.

But it was not all bad news. Far from it. I was really surprised by the change in Abel's own heart. It had enjoyed months of rest following the insertion of the HeartMate, and this had reversed his heart failure and changed his heart's globular shape back to normal. As we carefully dissected out the inflow cannula we found the bleeding point – a tear in the heart muscle itself. I peeled off the crescent of muscle attached to the metal inflow cannula and kept it for pathological examination, to compare it directly with the core of muscle excised to accommodate the inflow cannula in his first operation.

This was better than rocket science. We'd shown that the enlarged heart muscle cells had reverted to normal size and structure, and that we could help sick hearts to recover. We called this the 'Keep Your Own Heart' strategy. But were the structural changes sustainable and would the hearts continue to function? We didn't know. Only time would tell, but it was a monumental finding.

The surgery had taken seven hours. We delivered Abel's pump like a baby, as I wanted to keep it. The inflow cannula site was repaired with deep Teflon-buttressed stitches. His heart now looked like a dog's dinner, but it still worked and was contracting well, boosting his circulation as we re-warmed the blood. We separated from cardiopulmonary bypass as if it had been a simple, straightforward operation. There was bleeding from every cut surface but his blood pressure was fine.

Was this going to be the world's first successful 'bridge to recovery' in a chronic dilated cardiomyopathy patient? The bleeding eventually abated, and we closed the chest and abdomen. This was a triumph in itself. Abel's family were ecstatic, Ralph and Jean were relieved, and my staff were optimistic. But I was still unsettled. We were perched on a knife edge here.

I had no choice but to leave the post-operative care to the intensive care team. I was buggered, at best. At worst? Psychopathic, I guess, juggling too many balls in the air, pushing my own life – and other people's – to extremes. Surgery I find easy, politics less so. And taking risks with open-ended bills from the NHS was stressful. More than individual lives were at risk here. Many influential characters were claiming that mechanical hearts would never work and it was a battle to prove them wrong.

Abel remained completely stable for the next thirty hours, everything normal. His kidneys were passing urine despite the prolonged shock. Yet I was still uneasy. The stakes were too high, and I was walking on water but waiting to sink. I didn't have to wait long. Late at night Abel's own heart flipped into uncontrolled atrial fibrillation with a rate so fast that the left ventricle was suffering, a straightforward problem that happens to almost half of heart surgery patients. It should have been easy to fix, but it wasn't. None of the junior doctors on site dared shock him, so he deteriorated rapidly. I rushed into the hospital but by then he was beyond help.

Abel died with his family around his bed. I could do one of two things – go ballistic and get sacked, or walk away. I did the right thing, passing Ralph's bed on the way out. Jean was asleep, with her head on the sheets, oblivious to everything. Ralph stared straight forward, consumed with anxiety. His eyes followed me as I passed. He understood how I felt and there was nothing I could say to reassure him. He'd heard everything – 'Shall we shock him? Should we call the consultant? What if …?' Then the inevitable. Shit and derision.

There's such a narrow margin between life and death. Survival depends upon those present being able to treat the problem, upon whether the correct treatment is applied and if it's done at the right time. Abel needed that electric shock to return his fast heart rhythm back to normal. For that he needed someone to take charge and rescue the situation, but it didn't happen. It's what we now call 'failure to rescue'. I felt that he'd died needlessly, after all that effort.

Thankfully Ralph went from strength to strength. He'd been transformed by technology and soon learned to live with the 'alien' inside pumping away noisily, hissing through the air vent, circulating six litres of blood each minute with a strong, bounding pulse. Within two weeks both he and his family had mastered the equipment. The most important thing was dealing with the stiff white power cable as it emerged from his side. It needed to be kept scrupulously clean and the bugs kept out as the surrounding skin had to bond with it, integrating with the Dacron covering. Ralph's biggest risk was of drive-line infection, very common with this device and much worse for a diabetic like him. Indeed patients suffering from diabetes had initially been excluded from consideration for a heart pump for this very reason.

Jean practised dealing with unexpected problems and how to troubleshoot if the alarms went off. At such moments life itself depended upon doing the right thing, so she learned how to pump the HeartMate manually should the electrics fail. Then off they went, happy and confident, anticipating a new life, the swiftest hospital discharge of any artificial heart patient to date. Although Ralph came back for check-ups every month, they resumed travelling in their caravan, making the most of his resurrection. He was happy.

The winter brought predictable problems – a simple cold, coughs and sneezes. These caused shear stress at the stiff abdominal drive-line exit site, the delicate seal between skin cells and Dacron broke down, and bacteria infiltrated the break in the skin's defences. Jean tried hard to keep the area clean with normal drive-line care, but then it started to discharge pus, becoming hot, red and sore. Ralph's GP took a swab and put him on antibiotics. Infection made his diabetes more difficult to control, his higher blood sugar levels helping to feed the bacteria. After being on antibiotics for several weeks a fungus intervened, and we admitted Ralph to hospital for a few days to try to get on top of the problem. By now there was an infected and painful crater around the line, so we tried to revise it surgically. It certainly looked much better, and Ralph's own heart had improved considerably as he spent hours on an exercise bike building muscle.

Eventually the fungal infection reached the pump itself and I knew this was the writing on the wall. Over in Houston Bud was experiencing the same problem with his bridge in transplant patients, although none were diabetics, and I called him regularly for advice. We knew we could never sterilise it with antibiotics, but could we risk removing it as we'd done for Abel? I was seriously contemplating that when the infection gained entry to his bloodstream. Septicaemia, we call it. Now both the inside and outside of the pump were infected, the pig valves covered in masses of fungus and beginning to disintegrate. There was no way out of this. I had to explain to Jean that it was too late for heroics.

The septic shock had caused kidney and liver failure, and Ralph was now yellow, his lungs filling with fluid as the valves in the pump started to leak torrentially. The

HeartMate even sounded different, more like a washing machine as blood sploshed to and fro though the pumping chamber, its hiss now like a boiling kettle rather than a snake. For me it was finished, and Jean understood when I said it would be inappropriate to try 'Abel type' heroics. Ralph couldn't survive these. We should help his breathing with the ventilator and see him off with the dignity he deserved.

Ralph had helped to start something. 'The man with two hearts', as the *Sunday Times* called him, had done so well. He died eighteen months after the implant, surrounded by his family, and after all the suffering they remained grateful for this chance of life and time well spent.

We learned a considerable amount from Abel and Ralph. They were pioneers, the very first patients to receive an artificial heart on a 'lifetime basis'. We accepted that this 'lifetime' had been short, but all life is precious. Ask cancer patients about that. All we needed was better blood pumps – and we were working on that.

7

saving julie's heart

*'Ah, Nothing is too late, till the tired heart
shall cease to palpitate.'*

Henry Wadsworth Longfellow

WHY DID PATIENTS DIE after heart surgery? Was it because the surgeon made a mistake, damaging the heart through a technical error, operating on the wrong valve or coronary artery, or letting the patient bleed to death? Very rarely was it any of these. Usually it was because the patient was so sick beforehand that their survival remained in the balance even when the operation went well. As in any other profession mistakes could and did happen, but the majority of patients died because their diseased heart gradually deteriorated during the operation.

In conventional surgery at the time, the heart suffered during the period when it was deliberately stopped without its blood supply, irrespective of the protective solutions we infused, none of which were perfect. At the end of the operation it was simply too weak to sustain the circulation, tired yet potentially recoverable. When the bypass machine was turned down the heart just wouldn't take over, and without help the patient died on the operating table. Quite frequently the heart limped off the machine but gradually failed over the next few hours, and however much we flogged it with drugs, the die had already been cast in the operating theatre. The longer the heart muscle was deprived of blood flow, the more likely this was to happen. Then off to the mortuary the body went, leaving a grieving family behind.

I felt that this pathway to death was preventable. The heart just needed an opportunity to recover, and staying longer on cardiopulmonary bypass was not the answer. In fact it made things worse. The more time that blood interacted with the foreign surfaces, the greater the likelihood of whole-body inflammation, which in turn meant worse organ function and more bleeding.

So what about some other type of pump? A simple circuit without the oxygenator might work better, and this could be used for a few hours, perhaps days – or, in the worst cases, several weeks, until the heart's own contractile function and the benefits of the surgical repair would allow the circulation to be free-standing again.

A safe and reliable temporary blood pump would probably save half to two-thirds of those who might otherwise die. How did we know? Post-mortem examination showed us that the heart was structurally sound in most cases. It just got tired. Give it a rest and support the rest of the organs, then the patient might get better.

Inevitably most pioneers working on blood pumps thought that they needed to generate a pulse to replicate the human circulation. Early pumps had to empty and fill, and be large enough to mimic the normal heart. Usually it was just the left ventricle that needed help; if necessary, separate systems could be used to support both left and right ventricles. But the early pulsatile devices with bellows and valves created turbulence, friction and heat, a perfect environment in which to promote blood clot formation and the disastrous complication of a stroke – always a dismal and feared end point in the battle to save life.

At Allegheny General Hospital in Pittsburgh, George Magovern, the chief of surgery, was less convinced about the need for pulsatility. He argued that when blood reaches the tissues it's through tiny capillaries one cell thick. There's no pulse in this micro-environment as pulse pressure has already dissipated in the small arteries before reaching the capillaries. Should a pulse be unnecessary – as we'd suggested – then smaller, less traumatic pumps could be made, pumps that spin at high speed and deliver between five and ten litres of blood per minute. The pump just needed to be kind to the blood. So Magovern engaged his friend Professor Richard Clark, head of cardiac surgery research at the National Institutes of Health in Washington, DC, to work with him on the project.

It took the team five years to produce a spinning blood pump. It was the size of a bicycle bell and weighed just half a pound, electromagnets driving one single moving part – a six-bladed turbine. First called the AB-180, it was intended to support the circulation for up to six months, long enough for bridge to transplant. The design was so simple that one of the technicians attached a prototype to his garden hose and used it to drain his fish pond. It performed well in the laboratory without damaging the red blood cells and it worked fine when used in sheep. As a result the US Food and Drugs Administration (FDA) sanctioned a human trial with the AB-180 in 1997 on the strict understanding that the pump was only used on a 'last resort' basis. A trial of pump versus certain death.

In February 1998 I was invited to Washington by the FDA for a heart conference to discuss the recent operations I'd performed on Abel and Ralph. It was there that I met Richard Clark, who had been expected to retire but didn't want to cut the umbilical cord. Cardiac surgery was his life. Over dinner he showed me the AB-180 and asked whether I'd take him on as a research fellow for a year. I was flattered and suggested that he should bring the pump with him, and on 7 August that year he and his wife arrived in Oxford. It made for some stark contrasts: from skyscrapers to the dreaming spires, from the world's best-funded health care system to the National Health Service. Up until that point the AB-180 had still not been used successfully in a patient – three valiant attempts to rescue patients in shock all ending in death – and there was a distinct possibility that the clinical trial in the States would be stopped.

* * *

Two o'clock in the morning, 9 August 1998, and my phone woke me. Strange, as I wasn't on call that night. It was a cardiologist from the Middlesex Hospital in London. She was looking after Julie, a twenty-one-year-old student teacher who was home for the summer with her parents in Surrey and who'd initially complained of flu-like symptoms. Within days she'd become exhausted, listless and short of breath, sweating but cold, and not passing urine. Dying, in fact.

The district general hospital recognised this and passed her rapidly on to the London teaching hospital, where an ultrasound scan showed a poorly contracting heart. She had viral myocarditis – a viral illness like a cold but, when it involves the heart, potentially fatal. Inflammation and fluid accumulation had destroyed Julie's heart function, the cardiac output monitor confirming very poor blood flow throughout her body, less than a third of what it should have been. All in all it was a pretty desperate situation for a girl who'd been perfectly normal the previous week.

The cardiologist had admitted Julie to the cardiac intensive care unit for what we call a balloon pump. This is a sausage-shaped latex balloon attached by a catheter to an external air compressor, the catheter being fed through the leg artery up into the aorta in the chest and the balloon inflating when the heart relaxes. This raises the blood pressure and marginally reduces the amount of energy the heart needs to expend, but you do require some pressure and flow for it to work. In Julie it was bloody useless and just obstructed the blood flow to her leg. This was already blue – pouring out lactic acid – and at the time of the call the highest blood pressure was 60 mm Hg, half what it should have been.

I was considered the last chance saloon, and the Middlesex cardiologist wondered whether anything at all could be done. 'Is there any technology you've got that could help?' she asked, then reassured me that it was okay if I couldn't do anything as the shocked parents and younger sister had already said their goodbyes. They felt Julie had gone when she was anaesthetised to be put on the ventilator. Conventionally the ventilator and balloon pump were the final option – but they'd made no difference, and, inevitably, her blood pressure had dropped even further after the anaesthetic drugs.

Most patients with viral myocarditis get over it. As with influenza, the effects of the virus dissipate and the heart recovers – but this was not happening with Julie. The lethal blood chemistry and deteriorating organ function had gone too far, and she was bang in the middle of the vicious cycle of acute heart failure that invariably leads to death.

In the small hours of the morning you sometimes feel like saying, 'Sorry, I'm not on call. I've had a few beers and can't help you.' To be honest I don't recall what I said in this case, but it was probably along the lines of, 'Get her to Oxford as quickly as possible. I'll get the team ready.'

So Julie was brought by ambulance to Oxford in the middle of the night with doctors, nurses and masses of equipment. I called Richard Clark, who rushed straight in to unpack the kit, excited by the prospect of early action, and my earnest Japanese right-hand man Takahiro Katsumata came in to assist.

We met Julie and her helpers in the accident department following the harrowing sixty-mile dash from London. By then Julie's liver and kidneys had failed and her blood pressure was negligible, so we'd no choice but to rush her straight into theatre. She was as good as dead. Her parents hadn't arrived by then, still struggling to get out of London even at that time in the morning.

One thing that subsequent media reports stated was incorrect. They said that I'd been given the green light from my hospital's ethics committee to use the AB-180, but sadly this was wrong. Absolutely wrong. No one but me and Richard Clark had any idea that we had the device, and neither of us considered that we might need it so quickly. Up to this point it had a 100 per cent mortality rate, which is, to put it mildly, statistically significant. But I was not the kind of doctor who'd let a young patient die because of some bureaucratic detail.

It was fortunate that Brian, the perfusionist, had the heart–lung machine primed and ready. The intensive care doctor accompanying Julie already thought that they were too late, and when I put a hand on Julie's leg I too suspected she was dead. White and cold, her veins looked empty and her feet were blue. Even so it was difficult to move her quickly – although she didn't weigh much – and the drips, ventilator and balloon pump had to be shifted carefully. Katsumata and I lifted her gently onto the operating table, and Sister Linda was scrubbed up, gowned and set to go.

Dawn, the second nurse, stripped away Julie's white hospital robe. Her urinary catheter was caught in part of the equipment, stretched like catapult elastic, the inflated balloon still inside her bladder. Dawn fixed it. I told Linda to paint up with skin prep and get the drapes on. Katsumata and I scrubbed with haste – what was more important now, survival or sterility? Mike, our anaesthetist, grappled with the multiple lines and drugs, trying to make sense of it all, helped by the visiting anaesthetist, who held the key to the puzzle. It didn't really matter what went into the lines – nothing was helping. I asked Mike to focus the beam of the operating light on Julie's chest, then grabbed the scalpel.

The blade went straight through hard onto the bone at one stroke. Forget the electrocautery – we didn't need it. By now there was no circulation – so no bleeding from skin or fat – and Julie's heart rate was agonisingly slow. I ran the saw up the sternum. Again, no oozing from the bone marrow. We wedged in the retractor and swiftly slit open the pericardium with scissors. Mike pointed out that the ECG was slowing to a stop but I didn't need him to tell me that as I was watching Julie's swollen, virus-ridden heart. It just squirmed in a pathetic sort of way, like one of those toys whose battery is almost flat, the tin soldier beating his drum slower and slower till his arm finally stops in the air. Spent.

But while the heart was stopping, I kept moving. I stitched purse-string sutures in the aorta and the right atrium to hold the bypass tubes in place. The aorta was soft, with no pressure, the right atrium tense to bursting point. Every stitch hole pissed out dark blue blood carrying no oxygen. There was barely any blood flow to her lungs and by this stage I wondered whether she was retrievable.

Working like clockwork and saying nothing, we shoved in the cannulas to connect the bypass machine. Between each critical step I took Julie's buggered little ventricles into my fist and hand-pumped hard and rhythmically, like squeezing juice out of a grapefruit, a form of internal cardiac massage to maintain a semblance of blood flow to her brain and coronary arteries. That was all that mattered. Forget the guts and offal, just keep the brain and heart alive with whatever oxygen remained in the sticky blood.

Katsumata, a man of few words, murmured, 'Don't mention the war.' I told Brian to go onto bypass even before the venous drainage pipe was connected to the circuit, and almost black blood drained sluggishly into the tubing. In our haste we had an air lock in the drainage pipe from the right atrium, but this was no big deal. Lifting the tubing, air floated to the top, then, after the tube was dropped down onto the table, the air whizzed off into the reservoir.

Peace suddenly descended on the operating theatre as the once empty heart started to beat steadily, now that it was receiving blood from the machine. Julie's blood oxygen levels increased rapidly and her black blood turned red again, lactic acid filtering away. She was safe, as long as her brain was not damaged. A just-in-time job.

I turned to Richard. 'How do we implant this thing?' It seemed straightforward. There was an inflow tube, which I felt to be unreasonably rigid. This would be inserted into the left atrium to drain well-oxygenated blood from the lungs into the centrifugal pump. The pump would become her new left ventricle. There was a vascular graft to return blood to her aorta, which would then be circulated around the body. Simple. The device itself would sit in the right side of her chest between lung and heart. With the left side of the heart effectively bypassed, her brain and body would be safe. So let's get on with it.

Richard handed over the sterilised equipment to Sister Linda. I considered how best to insert the stiff inflow tube into the small, thin-walled chamber of the atrium. The entry point had to stay blood-tight for a long time so I thought we should sew a tube of human aorta to the left atrium. This would provide a degree of flexibility to the cannula entry site and make it safer to remove, without leaving a sizeable hole in the heart itself. This simple trick could make the difference between success and failure, life or death.

We kept donated human heart valves and bits of blood vessel in an operating theatre fridge for emergencies, and I had a special team whose job it was to arrange donations and rescue scraps from the autopsy room. These spare parts, pickled and preserved, were invaluable for congenital heart surgery, where we have to rebuild children's hearts.

Dawn found a suitable tube of donor aorta in a sterile bottle in the fridge. I carefully sewed this to an accessible part of Julie's left atrium and slid in the inflow cannula. It was all a bit Heath Robinson, making it up as we went along. Then, with careful, blood-tight stitching, I sewed the outflow graft of the AB-180 to the aorta using a side clamp. There was one last thing to do. The combined power cable and lubrication port needed to be passed out through a stab wound in the upper abdominal wall, making it look like we were wiring an android. I passed it to Richard and he connected it to the power supply.

By now, with the steady blood flow from the bypass machine, Julie's own heart was beating again. But it was still feeble. I decided that we should support her for another thirty minutes before attempting to switch from cardiopulmonary bypass to the AB-180 because, although the pump would take over from the inflamed and swollen left ventricle, the right ventricle had to look after itself. With this better blood flow the cut tissues started to bleed. What's more, she'd cooled as she was dying, and with the heat exchanger in the bypass machine her body temperature started to rise again.

I grew tired and a little impatient. I asked Mike to ventilate the lungs and Brian to leave some blood in the heart. We needed to fill Julie's own heart before switching on the AB-180, otherwise it would suck the heart empty and obstruct. We needed to slide imperceptibly from one to the other. But how? I told Brian to simply stop the bypass machine. He turned it off, and this confirmed that Julie's own heart was useless.

Then I told Richard to switch on the AB-180 and steadily turn up the flow to five litres per minute, equivalent to normal heart output. In a state of great excitement, he flipped the switch and turned it on. Immediately, the pump came to life. Julie now had bright red blood coursing around her body.

On the monitor there was no blood pressure trace – no systole or diastole – just flatline, continuous flow from the centrifugal blood pump. Would it work? We'd find out in the next few days. Until this point there had been a 100 per cent mortality rate in humans. But we could tell from the blood samples that it was looking good. Julie had pretty normal biochemistry. What's more, the homograft tube was working well. There was no bleeding around that crazy inflow tube, which had been a major problem in the three American patients. The turbine was spinning at 4,000 rpm, with a flow exceeding normal cardiac output, and the pump itself was perched comfortably on Julie's right diaphragm.

We'd succeeded in keeping her alive.

Somewhat perturbed by the flatline pressure trace, Mike asked Brian to switch the balloon pump on again. This produced a feeble pulse wave on the trace but absolutely no difference in blood flow to the body. But the pulse wave was much less important than blood flow. Every cell of the body needs well-oxygenated blood containing glucose, protein, fat, minerals and vitamins, and it really didn't matter whether the blood had a pulse or no pulse in it. Flow was the key.

This was a complete revelation at the time. Systole and diastole had always been considered so important, and you had to continually measure them. If blood pressure was low you had to get it up. But this was not the case with a continuous-flow pump. Low blood pressure actually provided less resistance for the pump to work against. When pressure went up pump flow went down. Counterintuitive physiology. We had to get used to it.

It was almost 8 am and the sun shone brightly on the dreaming spires. I left Katsumata to close the chest and went to warn the intensive care unit about the impending arrival. It would be something completely different for them. I told them that for the next twelve hours – Julie's critical period – she'd have no pulse, and that an average blood pressure of 70 mm Hg was fine. Her kidneys had packed up, so she'd need dialysis for a few days. And she was a little yellow, as her liver was suffering as well. In fact, when she'd arrived in the ambulance from London by most criteria she'd been dead. But we hoped she wouldn't be dead now. Good or what?

Desiree Robson, our chief nurse, asked whether I'd talked to the family. They were sitting in the relatives' room – Mum, Dad and little sister, totally exhausted after their night-time chase around the south of England – awash with tea and sympathy, but still expecting bad news.

'Go and tell them what's going on,' Sister ordered. 'Celebrate later.'

At that point I was unsure what I could tell them. Try this – 'Your precious daughter arrived too late. We all thought she was dead despite the ventilator and balloon pump, but we implanted an unlicensed, previously completely unsuccessful machine from the States. And now we have resurrected her from the dead. As long as her brain still works, that is.' This was the harsh truth of it all.

I walked into the miserable relatives' room, where the clock was still stuck on 5. Three heads were bowed, hands wrung in laps. They all looked up at once, and I could tell right then that even though they'd no idea who I was they just knew I was there to tell them the worst. Then they read my expression. With mask dangling down and blood on my theatre shoes, I looked pleased, and not with the sycophantic, forced look of sympathy doctors put on when giving bad news. Julie was still alive, a miracle of science.

I didn't explain that it was new, untested technology that had never before succeeded. The nurse allocated to Julie's ITU bed slipped in behind me, quite appropriately, to hear what I would tell them. But nurses hate it when I suggest that everything will be fine. They want me to look grave and talk about a critical period just in case something goes wrong. They don't want me to put the unit under too much pressure. Pressure to get things right.

All I could tell them was that the pump we'd used was keeping her alive and we'd been very lucky. It had only arrived from the States two days before, and we'd unpacked it with Julie already on the heart–lung machine.

'What are her chances now?' Julie's mother asked.

I told her that we hoped it would keep her alive until we could arrange a transplant. We weren't a transplant centre but I would talk to one and make it happen. It wasn't the time to mention that I was scheduled to be in Japan in three days' time.

I left the relatives there in the room. I was told that Mike and Katsumata were bringing Julie around, and that her mum and dad would be able to see her soon. Although it might be distressing for them – there were many tubes and lots of equipment attached to her diminutive body – it was better than visiting her on a slab in the morgue, with ashen white face and waxy cold hands, lips bruised from the tracheal tube. I knew well from experience that anything was preferable to that.

Sister Desiree was there to sort things out – unravel the drips, plug in the machines, calibrate the monitors. Get it all absolutely right. Desiree and Katsumata would be experts on the AB-180 by the end of the morning. For now they had to get used to looking after the girl with no pulse. This team didn't need me, which was just as well. My mobile phone rang. The signal was poor but the message was just discernible – the medical director wanted me to come to his office.

I was expecting this, and knew that I wasn't being invited round for coffee. Medical directors are the Stasi from a hospital doctor's standpoint. Put simply, they are there to ensure that no one does anything new or interesting. Anything that might generate bad press for the hospital. As they say in court, I had previous form. A loose cannon.

His face was like thunder. How dare I use an unregulated device? Who knew about this? Were the ethics committee involved? What on earth was I trying to do, keeping this young girl alive? He didn't say any of this, but that's how it came across.

I didn't respond, but just sat there in my blood-stained theatre gear thinking, 'Get a life.' It was time to play the obvious card. I said I didn't have time for this and needed to get back to the patient. His parting comment was, 'If you do anything like this again you'll be out.' This reminded me of repeated threats to send me to a bad boys' boarding school as a child. They never worked.

I went straight back to the ITU. Julie's family were now by the bedside and Desiree was explaining the paraphernalia keeping her alive – breathing machine, balloon pump driver, AB-180 console, infusion pumps, warming blanket. All quite simple, really. And they were bringing in the dialysis machine for her kidneys. By now the operating theatres were waiting to start the day's planned cases. I told them I was ready and that they should send for the first patient, a premature baby with a big hole in the heart whose parents were getting anxious.

Between operations I kept going back to Julie. I couldn't see the bed for doctors. One of my cardiology colleagues was trying to get good ultrasound pictures of Julie's heart without interference from the adjacent pump, and these were provoking great interest. The ventricular muscle was completely offloaded and doing no work, well and truly rested, and only a slight twitch remained to show that electrical activity continued. The flat line on the monitor unnerved some of the medical staff.

By early evening everything was stable and the crowds had melted away. With an empty left ventricle and low blood pressure, the balloon pump was superfluous. Not only that, it was partly blocking Julie's leg artery and was just another route for bacteria to gain entry to her system. I insisted they remove it. Katsumata lived in the hospital complex, Desiree just a couple of streets away. They said they'd keep a close eye on her, so I set off home for the night, away from the madhouse.

By early morning Julie was awake. With the breathing tube down her throat she was frightened and agitated. She had no idea of her whereabouts or why she had apparatus emerging from every orifice of her body. And she was clearly in pain, so we needed to sedate her again. Just enough, because too much would drop her blood pressure. An injection of barbiturate into the drip and she drifted away again into oblivion, the best place to be in these circumstances.

I put a stethoscope over her sternum and heard the loud, continuous whirring sound of the magnetically suspended turbine – still set at 4,000 rpm – pumping five litres per minute, the same volume pumped by a normal heart. Few of the people at her bedside, on the ITU, in the hospital, in Oxford – or even in the country as a whole – realised the significance of this one, single case. Pulseless flow was causing Julie's organs to recover – brain, kidneys, then liver. The pioneers of artificial heart technology had denied that this was possible, claiming that pulsatile pumps were essential and blaming the three previous failures with the AB-180 on this fact.

So what was the significance of this finding and why was I starting to get excited? If pulseless flow worked this well temporarily, then the new Jarvik Heart should be successful in the longer term.

At 7 am I was called to the phone at the nurse's station. Someone with an American accent wanted to speak to me – they didn't get the name. It was George Magovern, the man who'd initiated the AB-180 project, calling from Pittsburgh well after midnight local time. Richard had called him but he wanted to thank me personally. His engineering team were still out celebrating and they wished Julie luck, hoping that we could keep her going till a donor heart became available. I said we'd try. This was just the boost I needed right then, something to put the sceptics in perspective. And the medical director.

The next day we took her off the ventilator and removed the tracheal tube. Miraculously her brain seemed fine. She could talk to her parents and there was more urine in the bag. I watched the flat line on the monitor screen. Then I noticed something. Her regular heart rhythm had changed to irregular atrial fibrillation. This was not unusual in itself, but when there was a long pause after irregular beats a definite blip appeared on the arterial trace – her own heart was starting to eject blood when allowed to fill for long enough.

I didn't say anything but wondered whether her heart was beginning to get better. Most cases of viral myocarditis improve with medical treatment before they ever reach the shock stage. So why would we want to give Julie a transplant if her own heart was recovering? It was simply the conventional thing to do with severe heart failure. I suggested that we give her a dose of steroids to help

decrease the swelling in the muscle. Witchcraft, but if
nothing else it would make her feel less lousy.

I now had a very difficult decision to make. Today was
Wednesday. Through some curious oversight I was scheduled to talk at a conference in Japan on Friday and
another in South Africa on Saturday. Unbelievable planning. Clearly the dates had been written in the diary as if
they were London and Birmingham, yet it was just about
doable. The question I asked myself was whether I should
go at all. With the time differences I even had difficulty
working out how long I'd be away. But no one's indispen-
sable, I had a great team and Julie was stable. So I decided
to go.

Before I set off we had a team meeting – Katsumata,
Richard, Desiree and the intensive care doctors – as we
needed a plan for the time I'd be away. The signs were
good: Julie's kidneys and liver were already recovering,
there were regular blips on the arterial pressure trace and
echocardiogram pictures showed improvements in her
heart muscle contractility. The pump was doing its job.
The plan was to keep her stable and just let her recover
slowly. This required a steady nerve.

A few days later I received the sort of message I dread.
When I switched on my mobile phone in Johannesburg
airport on Saturday there was a worrying text from
Katsumata. They thought Julie was bleeding into her
stomach, a common stress response but made worse by
anticoagulation for the pump. But. The big but. Her own
heart was much better on the echo pictures. With the
pump turned down the left ventricle generated virtually

normal blood pressure. I wondered whether the steroids had helped the heart but caused the gastric bleed. I needed to talk around the situation.

I sent Katsumata a text. 'Now in South Africa. Ring me.'

His call followed soon afterwards. 'How was Japan?' he asked.

'Great,' I replied. 'Just don't mention the war.' Then my punchline – 'Don't stop the anticoagulation yet. Turn the pump down to 1,000 rpm for an hour. If the heart still performs well, take the pump out.'

Then came a long pause. I could sense Katsumata's 'Oh shit' moment. The silence dragged on until I said, 'Come on, Katsu. You and Richard can do it. Just get the bloody thing out.'

Katsumata had called me from Oxford early on the Saturday morning. He returned to the bedside with Richard and they called for another echo. Reducing the pump speed allowed the left ventricle to fill and eject more blood. They asked Julie whether she felt any different and she told them she felt fine. She just wanted it out. She'd experienced no more breathlessness and there was still a normal blood pressure trace on the screen. Richard knew that the lower the speed, the higher the risk of clotting in the pump or vascular graft.

Desiree was starting a blood transfusion and asked what I'd said on the phone.

'He told me to take the pump out and don't mention the war,' Katsumata said with trepidation. 'And one last thing. Only let the medical director's office know after it's out. We don't want him to have a stroke.'

'Then you'd better inform theatre and get on with it,' Desiree replied.

Richard and Katsumata explained the balance of risks to Julie and her parents. If her heart had already recovered but she bled to death from her stomach it would be a disaster. Even Richard, with his vast experience from his role in Washington, had butterflies. For him the stakes were high, as at last he was on the verge of success with the AB-180. But it was Julie's life that really mattered.

So Katsumata took Julie back to theatre seven days after the implant, ironically about the usual amount of time for a viral illness to get better. Richard didn't have hospital clearance to operate so he just had to watch, although had there been an issue he'd have dived in. He had no problem with this, filled with cautious optimism at the prospect of success.

Julie's heart was looking good – the stiffness and swelling gone, her blood pressure stable and receiving just a little help with drugs in the background. They still had a balloon pump in reserve but she didn't need it. Katsumata washed out the whole chest with warm saline solution, assiduously removing old blood clots from the chest cavity and from the pericardium surrounding her small but enthusiastic heart. He inserted clean chest drains, then closed the sternum tightly with wire. For the last time.

It was important to preserve the forward momentum. Julie woke up quickly and felt much better off without the breathing machine. The tracheal tube was removed later that evening, Desiree simply ignoring her shifts and staying with her, all the while encouraging her to breathe deeply and cough despite the pain. The anticoagulation had stopped, and the blood loss from superficial stomach erosions ceased soon afterwards.

We'd done it. We'd saved Julie's heart.

When Katsumata rang me with the news I'd already given my talk and was back at Johannesburg airport, homeward bound, relieved and in the mood to celebrate. Then Richard called George Magovern and his team in Pittsburgh, spreading the joy. But none of us were happier than Julie's family, their grief and desolation lifted, and suddenly with no funeral to plan. One day soon they'd take her home, Oxford just a grim memory.

In the 1990s any patient who received a left ventricular assist device in the United States was committed to a cardiac transplant, and few other countries had access to circulatory assist technology. What we achieved with Julie became known as 'bridge to recovery' in contrast to the conventional term 'bridge to transplant'. The procedure had not been done before in the UK, and bridge to recovery – our 'Keep Your Own Heart' strategy – soon emerged as the preferred approach for critically ill viral myocarditis patients. I was proud of that.

Just before Christmas 1998 the Pittsburgh engineers and researchers who'd worked on the AB-180 filed into a conference room for a special party arranged by Dr Magovern. No one knew what the occasion was – until Julie walked in with her sister. 'The girl without a pulse' was instantly recognisable from photographs that had been pinned to bulletin boards at the time of her groundbreaking operation, her face destined to grace the front pages. There was a moment of stunned silence, then loud cheers. Julie blushed as George shook her hand.

'You being here is the best Christmas present any of us could have,' he said.

He was right. The company survived and thrived, and the AB-180 was modified so it could be used without opening the chest. Now called the 'Tandem Heart', it's used worldwide to support patients with shock in the cardiac catheterisation laboratory.

Julie remains well almost twenty years later, and she works in a hospital. I look forward to a reassuring card from her family every Christmas. Long may her good health last.

8

the black banana

'We shall never surrender'

Winston Churchill, during the
Battle of Britain, 1940

MONDAY 15 FEBRUARY 1999, 3.45 am. No one calls with good news at night. I'd been in Australia for just thirteen hours after a twenty-hour flight. In pitch darkness I scrambled across my hotel bed and knocked the receiver to the floor. The call was lost. I swiftly slipped back into sleep, courtesy of melatonin tablets and the bottle of Merlot I'd drunk at dinner. Ten minutes later the phone rang again. This time I fielded it successfully, but I was irritated.

'Westaby? This is Archer. Where are you?'

Nick Archer was the consultant paediatric cardiologist at Oxford.

'Nick, you know I'm damned well in Australia. It's the middle of the fucking night – what's the problem?'

I didn't want to hear the answer.

'Steve, I'm sorry, but we need you to come back. We have a sick baby with ALCAPA. The parents know you and want you to do the surgery.'

Oh joy.

'When?'

'As soon as possible. She has bad heart failure and is as good as we can make her. The ventricle is poor.'

Already there was no point in further discussion. I pictured the frantic parents, desperate for an operation before it was too late, and the four grandparents huddled around the cot trying to lend support but only transmitting anxiety. I really had no other choice.

'OK, I'll fly back today. You tell the team we'll do it tomorrow, whenever that is.'

It was high summer in the southern hemisphere and early-morning light had just begun to penetrate the drapes. Further attempts to sleep were pointless. I stepped through the curtains onto the balcony and gazed out at arguably the finest city vista in the world. Across the harbour the first hint of sunrise cast ghostly shadows over the Opera House. Flags fluttered on masts in the harbour below, and to my right the tall, white city lights stood out against the pink morning sky. The peace was broken by the shifting gears of a Harley-Davidson. Maybe a surgeon racing into Sydney.

In Oxford a real-life tragedy was unfolding for this little family. Kirsty was a beautiful six-month-old baby girl in whom fate had installed a lethal self-destruct mechanism, a miserable detail that seemed destined to end her life before her first birthday. ALCAPA is an abbreviation for Anomalous Left Coronary Artery from the Pulmonary Artery, an isolated and exceptionally rare congenital anomaly in the overall complexity of human anatomy.

Simply put, it's bad wiring. Both coronary arteries should rise out of the aorta and supply the heart muscle with well-oxygenated blood under high pressure. They should never be attached to the pulmonary artery, as this has both low pressure and poor oxygen content. Early survival with ALCAPA therefore depends on the development of new 'collateral' blood vessels between the normal right coronary artery and the misplaced left coronary artery. But these eventually are insufficient to sustain blood flow to the main pumping chamber. Muscle cells deprived of oxygen die and are replaced by scar tissue, leaving baby to suffer what are in effect repeated, painful heart attacks. Scar tissue stretches, causing the left ventricle to dilate, then the heart progressively fails and the lungs become congested with blood, leading to breathlessness and exhaustion. Even during feeding.

So by six months of age Kirsty already had the same problem as my grandfather – heart failure through end stage coronary heart disease. But because ALCAPA is exceedingly rare, the diagnosis is seldom made until the infant is terminally ill. Fortunately, her parents were intelligent, had recognised that there was a serious problem and were persistent in finding help for her.

Kirsty's story was particularly harrowing. Her mother Becky already had a three-year-old son and was an experienced and responsible mum. She'd contracted no illnesses, smoked no cigarettes or drunk any alcohol during pregnancy, risking nothing that could potentially harm the foetus inside her. All antenatal checks and ultrasound scans appeared normal. Kirsty was born on 21 August 1998 by elective Caesarean section with a spinal anaesthetic and at first everything seemed normal. But not for long.

In the womb the pressures and oxygen content of the aorta and pulmonary artery are the same, so Kirsty's tiny heart was safe. After birth the circulations to the body and newly expanded lungs separate, and both pressure and oxygen content in the pulmonary artery fall. So, in cases of ALCAPA, both blood flow and oxygen content in that critically important left coronary artery fall precipitously, too. Kirsty was grunting even during the first attempt to breastfeed in the hospital, and Becky noticed beads of sweat trickling from the bridge of her baby's nose. The effort of feeding repeatedly made her fractious and distressed.

This was distinctly different from how her son had been, so Becky asked for a paediatrician to review Kirsty. She was told that there was absolutely nothing to worry about. This is exactly what anxious parents want to hear, but the truth was that no one had bothered to find out what was wrong. It was too much trouble – piss-poor medicine. At this stage Becky had little choice but to take her irritable but precious little bundle home.

Within weeks Becky was certain that something was seriously amiss, because during every feed there was sweating and vomiting. Kirsty struggled for breath, clenched her little fists and screamed until she was puce in the face. Together they made many visits to the GP, sometimes as many as three times a week, but they always received the same non-committal reassurance – tense, unpleasant encounters, as Becky was deemed neurotic and unable to cope.

But despite Kirsty's rapid breathing she had no temperature. Chest infections were ruled out, and her belly was soft, with no signs of stomach or intestinal blockage. All of the common paediatric problems were excluded. Family and friends offered rational explanations – it must be colic and would get better. But with her husband working abroad Becky became more and more anxious. Kirsty wasn't gaining weight. She had a pasty, washed-out look and a cough like a dog's bark.

In reality this baby was suffering repeated small heart attacks with excruciating chest pain that she could neither communicate nor understand. The human body can be outlandishly cruel.

Eventually, after a meltdown in the GP's surgery, Becky insisted that Kirsty should be referred to the local hospital. Twice she had chest X-rays, only to be diagnosed with bronchiolitis – inflamed breathing tubes – on both occasions. Then one day during her afternoon nap Kirsty turned a terrible slate-grey colour. She was barely rousable and limp. In abject panic Becky snatched her up and rushed to the surgery. But by the time they presented themselves at the receptionist the baby was awake and pink again. There followed yet another put-down. Becky was told to stop fussing and that there were children to be seen who were actually sick. On this occasion mother and child were acrimoniously dispatched, once more with a prescription for antibiotics. Kirsty's enormous heart remained undetected.

Becky's anxiety and frustration now turned to desperation. Every instinct told her that if she didn't push further something dreadful was going to happen, so she drove directly to the accident and emergency department of her small local hospital. They were seen by a sympathetic female doctor who had children of her own. Recognising that a mother's instinct was usually correct, she referred them on to a larger city hospital for review by the on-call paediatrician.

It was a bitterly cold, frosty night and they were left sitting in an unheated hospital corridor for several hours. Becky frantically struggled to keep Kirsty warm but she became progressively more limp and grey. Eventually, late into the night, they were seen. The first junior doctor suggested bronchiolitis and intended to dismiss them without investigation. Bronchiolitis seemed to be the only paediatric diagnosis these doctors had heard of. Becky was angry and frustrated, yet afraid that she would be thrown out if she protested.

When she refused to leave without a chest X-ray she was told off for her unreasonable attitude – how thoughtless to inconvenience the hard-pressed duty radiographer at that time of night. So the sad pair were dispatched unescorted down poorly lit corridors and icy outside walkways to find their own way to the X-ray department. It was well past midnight when they returned, clutching the tell-tale picture, which Becky presented to a nurse. They were parked once more.

Another thirty minutes passed, then a dramatic shift in attitude from the hospital staff. Kirsty and Becky were ushered into a cubicle with different doctors. Now there were hushed voices, grave expressions, and nurses bringing drips and drugs. This was even more frightening than being ignored. The previously mean, now embarrassed nurse took Becky aside to explain that Kirsty was being transferred to the specialist children's heart unit in Oxford. By ambulance this time. Suddenly she was too sick to remain unsupervised.

So what did the X-ray show to trigger this frenzy of activity? Kirsty had a massive heart. No one before then had bothered to examine her, but her problem was immediately obvious on the X-ray film. When the staff were pressed about her previous X-rays at the same hospital, Becky was told that the heart shadow had been misinterpreted as fluid – 'Sorry, but it's an easy mistake to make.' Some mistake! How do you begin to describe a mother's anxiety that hits like an axe, drains blood from the throat and takes your legs away?

Things were different when they arrived in Oxford. The paediatric cardiology registrar met the ambulance and took them directly to a ward packed with children with serious heart problems and beeping monitors – a hive of activity in the depths of night.

Nick Archer arrived at 3 am. On examining Kirsty he was immediately concerned by her body temperature. Despite Becky's best efforts she was cold and needed to be in an incubator. An ECG and blood tests were done quickly, then the echo machine was brought in to image Kirsty's heart chambers. First it seemed like good news. All four chambers were there, with no holes between them. But worryingly the left atrium and ventricle were both enlarged, the ventricle dramatically so. This explained the heart failure and accounted for the striking chest X-ray.

In little more than an hour the cardiology team established that Kirsty had severe heart failure from multiple heart attacks. Parts of the left ventricular wall now consisted of thin scar tissue interspersed with poorly contracting muscle, a rare condition in an infant but one that provided the likely diagnosis. One further test was needed. A cardiac catheter would confirm the diagnosis but would require a general anaesthetic, so she'd have to be in much better condition before proceeding.

By now, waiting in the hospital, Becky was bereft with grief, physically and emotionally drained. Her husband was away in the States on business and she felt very much alone. Guilt and irrational thoughts preoccupied her. Had she exercised too much during pregnancy? Drunk too much coffee? Offended God? There has to be a reason for everything. Deep despair took a grip, and her anxiety soon progressed to outright panic. She was certain she'd lose Kirsty. But as the winter sun crested the horizon she lost consciousness for a couple of hours. When she awoke the ward was busy, full of positive, warm-hearted people who tried to reassure her that, even though things were difficult, there was a great team caring for Kirsty.

It was a full five weeks before Kirsty was considered fit enough for the cardiac catheter. Becky had her husband back to share the pain, and the evening before the operation Mike the anaesthetist came round to talk. Normally a jolly and optimistic character, on this occasion Mike didn't have much to smile about. He warned the family that Kirsty's heart was so severely damaged that they might lose her during the procedure. It was only fair to inform them of this. So that night Kirsty was christened in her cot by the hospital chaplain, the doctors, nurses and other families gathering around the bed to support them.

Everyone knew what the catheter would show. There was only one rare condition that did this to a baby, that caused multiple heart attacks in the first few months of life – ALCAPA. Becky overheard the words 'early surgery' and hoped that they didn't mean a transplant. Both she and her husband stayed by the cot the whole night long, terrified that Kirsty might slip away. In the morning, without having slept and paralysed by fear, Becky dressed her baby in her best pyjamas and tied a bow in her hair for her trip to the catheterisation laboratory. Ironically it was Valentine's Day. As Becky put it to me later, 'A girl has to look her best, even when she's poorly.'

Once I was on the plane home I started to sketch the anatomy of Kirsty's aorta, pulmonary artery and abnormal left coronary artery. I knew that the current techniques in operations for ALCAPA had limitations and a substantial failure rate, so I used my time during the flight to work out an alternative. By the time we were cruising high above Java I'd designed my new operation. Last on,

I was first off the plane back in London. As I waited for the airbridge to connect and the doors to open, the cabin services director handed me a bottle of champagne, wished me luck and whispered, 'You operated on my sister's baby.' Small world. I thanked her.

I called my colleague Katsumata when I got back to Oxford and asked him to bring Kirsty's parents to my office with a consent form. The cardiac catheter had shown precisely what Archer had suspected. Kirsty needed surgery as soon as possible.

Becky looked tired and drawn when I saw her. She knew instinctively who I was. Her face lit up as they entered my office in a cold Portakabin.

'We're so pleased to see you,' she said. 'How was the trip?'

'Good. Very restful,' I lied. 'We need to get on with this, don't we?'

Katsumata had managed to find an electric convection heater to take the chill from the room, and we all set about breaking the ice. They explained that a family member was a representative for a heart valve company and knew me well. He'd been expecting to see me at the meeting in Australia. They were sorry about my aborted trip but profoundly grateful that I'd come back as they wouldn't let anyone else operate on their little girl. Despite the warmth, Becky was now trembling uncontrollably with abject fear. Poor kid. Finally, after weeks in hospital the time had come – the day she might lose her baby.

I don't do transmitted anxiety if I can help it. But it's much more difficult for my anaesthetist colleagues who have to deal with the agonising separation when the patient is given over to them. I described my planned operation to the team and explained why I felt it would be an improvement on existing techniques. The new left coronary artery would be constructed with a flap of aortic wall that would sit below a corresponding pulmonary flap to make a tube, the latter containing the misplaced origin of the left coronary artery at its apex. The product would be a new coronary artery delivering high-pressure, well-oxygenated blood directly from the aorta, where it should have come from in the first place. Blood fully saturated with oxygen would then supply the failing heart muscle and prevent further heart attacks. Katsumata was intrigued and excited by my proposed approach, so much so that he rushed off to call the hospital's film crew.

With severe heart failure, the risks of the operation were substantial. Becky's shaky hand signed the consent form and I walked back with them to the children's ward. When we reached the cot Kirsty's heart failure was worse than I imagined, indeed the worst I'd ever seen in any child. She was emaciated, with virtually no body fat, her heaving ribs and rapid breathing a consequence of her congested lungs, and her abdomen swollen with fluid. She was still a pretty baby, but without immediate surgery she'd be dead within days. Although a voice in my head screamed, 'Oh shit,' my mouth correctly said, 'I'll go to theatre now.'

Mike and the nurses were busily preparing drugs and catheters in the anaesthetic room. He knew the score, having already anaesthetised Kirsty for the cardiac catheterisation, and some of the monitoring lines were still in place.

'Do you really think you can get this baby through?' was his opening line.

I didn't reply, bidding a cheery 'Good morning' to the nurses and perfusion team in the operating theatre, then went straight to the coffee room. I wanted to avoid seeing Becky leave her baby with strangers, always an excruciating event.

When I returned, Kirsty was already on the operating table, covered in green drapes held in place with an adherent plastic drape. All that was visible were her bony little chest and swollen abdomen. Heart surgery needs to be an impersonal, technical exercise.

I joined Katsumata and my six-foot, six-inch Australian colleague Matthew at the scrub sink. While we scrubbed in silence the film camera was carefully positioned next to the operating lights. There was a palpable buzz of excitement. We were about to do something novel, esoteric and risky.

There was no bleeding as I drew the scalpel blade along the skin over Kirsty's breastbone. In shock, her skin capillaries had shut down to divert blood to vital organs. Next the electrocautery cut through the thin layer of fat onto the bone, producing its characteristic buzzing noise accompanied by a whiff of burning as the current cauterised the oozing blood vessels, although this time there were few of these. Then the electric saw cut through the length of her sternum, exposing bright red bone marrow.

We used a small metal retractor to crank open her tiny chest, bending and stretching the joints between the ribs and the spinal column. In babies the fleshy thymus gland lies between the sternum and the fibrous sac around the heart, but by now it had done its work producing antibodies for the foetus so we removed it.

The electrocautery continued its messy but vital work, cutting through the fibrous pericardial sac to expose the heart, straw-coloured fluid pouring out and being drawn away by the sucker. Meanwhile the other members of the team worked on silently. Mike gave heparin to stop Kirsty's blood from clotting in the heart–lung machine, the perfusion team set up the complex array of tubing, pumps and oxygenating equipment to keep Kirsty's body alive when her heart was stopped, and the scrub nurse Pauline concentrated on having the correct surgical instruments ready to slap into my palm. I rarely had to ask for anything. This complex, highly coordinated work relies heavily on having a steady, consistent team, and as most of them had been with me for years they enjoyed my complete confidence.

As we pulled up the edges of the pericardial membrane to display the heart, Katsumata audibly drew breath and murmured, 'Oh shit.' It was truly a frightening sight. Back from his first cigarette, Mike popped his head over the drapes in response to Katsumata's comment. I agreed that things were even worse than we'd thought. Others could see it all on the video screen.

What should have been a walnut-sized heart was revealed to be the size of a lemon. The enlarged right coronary artery was obvious, its many dilated branches crossing over towards the left ventricle. While the right side of the heart pumped vigorously against raised pressure in the lungs, the left ventricle was hugely dilated and barely moved. Patches of newly necrotic muscle merged with areas of white, fibrous scar tissue, the result of many small, painful heart attacks endured by Kirsty during her first six months of life. Katsumata was right to be concerned, but I didn't respond to his anxieties. We were committed to rectifying that blood supply and hopefully improving matters. Kirsty had survived to this point and it was our job to keep it that way.

Having exposed this heart I began to question the wisdom of attempting such a complex operation having come straight from a day-long flight. Yet what would have been gained by turning her down for surgery or procrastinating further?

For Kirsty there was no alternative. Urgent heart transplants were virtually impossible in babies, so this replumbing of her heart's blood supply was her only chance of life. Grim Reaper was perching on the video camera and we all knew it, but now I was committed there could be no turning back.

Tiny pipes were inserted to connect her to the heart–lung machine, and I then gave the signal to go on bypass. The perfusion technician turned on the roller pump and Kirsty's heart gradually emptied. At this point technology had taken over, diverting blood away from her lungs and into the synthetic oxygenator. With her empty heart still beating I cut through the pulmonary artery above the origin of the anomalous coronary. There was the opening to the vessel, like the pearl in an oyster. Now we had to connect it without tension to the high-pressure aorta that lay almost an inch away. The conventional method was simply to try to stretch and re-implant the origin of the vessel into the side of the aorta. But this could result in thrombosis and blockage, so I pressed on with my new technique.

This delicate exercise could only be achieved by clamping the aorta and temporarily stopping all blood flow to the heart. We'd protect the muscle by infusing cardioplegia fluid directly into both coronary arteries, flushing all the blood out and collapsing the ventricle like a punctured football. This induced state of inactivity, common in heart surgery, is reversed simply by removing the clamp on the aorta, which allows blood from the heart–lung machine to flow back into the coronary arteries.

For the reconstruction of this tiny vessel the stitching had to be precise, accurate and watertight. The procedure went well. Just thirty minutes after the heart was stopped, the conjoined flaps restored Kirsty's coronary artery anatomy to what it should have been. As the clamp was removed, bright red oxygenated blood – rather than deoxygenated blue blood – flooded the left ventricular muscle. Her heart changed from a pale pink colour to deep purple, then became almost black in parts. Before reconstructing the pulmonary artery we checked that there was no bleeding from the lines of stitching behind it. Soon the electrocardiogram showed uncoordinated electrical activity, and the heart stiffened with renewed muscle tone.

Unusually for a child, her reperfused heart kept writhing and squirming in ventricular fibrillation. We used an electric shock directly through the muscle to restore normal rhythm. Ten joules – zap! The heart defibrillated and stopped wriggling. It was now motionless but we expected a normal rhythm to resume at any moment. But it didn't. The purple ball fibrillated and squirmed again, and the anaesthetist's head popped over the drapes to request the obvious – 'Shock it again!' We did and the same thing happened. It wasn't coming back.

This was serious electrical instability caused by the scar tissue, so we gave the appropriate drugs to stabilise the muscle cell membranes.

'Let's give it more reperfusion time,' I told Mike.

'OK, I'll go out for a fag then,' he said.

Twenty minutes later we tried again. Twenty joules – zap! This time her whole little body levitated from the operating table and her heart defibrillated. Although it slowly began to beat, it was barely more than a flicker. Ominous, but we had drugs in reserve to make it pump harder.

I asked Mike to start an adrenaline infusion and told the perfusionist to cut back on pump flow to leave some blood in the heart. This was operating theatre protocol, and it's just like the military. You make a request to a medical colleague but give orders to the technical staff. If you start giving orders to an anaesthetist they'll tell you to piss off, and will go off and do something different.

While Mike and the perfusionist worked together to check and optimise the blood chemistry, my gaze remained fixed on Kirsty's pathetic little heart. The new coronary artery was fine – there was no kink in the tube and no bleeding. For the first time the left ventricle was receiving well-oxygenated blood at the same pressure as the rest of the body. But her heart still looked like an overripe plum and was barely beating at all. Moreover, the mitral valve was leaking badly. Although I heard myself telling the team to give it another half hour's support on the pump, what I was really thinking was we're stuffed, this heart's had it; great operation – dead baby.

Of course, I didn't let the others know my thoughts. They'd salvaged so many catastrophes that they expected me to pull this one off too. But I was starting to fade. I suggested that the cameraman should stop filming for a while because nothing was going to change and asked Katsumata to come to my side of the operating table while I took a break. I removed my gown and gloves, and went to make a call in the anaesthetic room. Mike followed.

'Can you repair the mitral valve?' he asked me.

'Don't think so,' I replied. 'I'll get Archer to warn the parents.'

I slumped on a stool and picked up the phone. One of the lovely nurses put a coffee and doughnut in front of me. With her arm around my shoulder, she felt the cold sweat dripping down the nape of my neck.

'I'll get you a dry top,' she said.

In five minutes Archer was down from his outpatient clinic at the theatre door. He didn't need to ask.

'Thought you might have trouble. Anything I can do?'

'Take a look at the echo,' I said. 'The repair's fine but

the ventricle's lousy. Mitral valve's leaking. At this rate we're not going to get off the pump.'

My bladder was full and I wandered off to the loo. When I got back, my brain had regained control without that distraction and I now really needed to focus. What, if anything, could I do to make things better? I was running out of ideas.

The left ventricle was scarred, dilated and now globular – not the normal elliptical shape. This distortion had pulled open the mitral valve and prevented it from closing. As the left ventricle tried to pump blood around the body as much as half of it flowed backwards to the lungs. Heart function is always temporarily worsened during surgery, but in Kirsty's case it seemed terminal. I'd hoped that resting the heart on the bypass machine would help it to recover. It hadn't.

I went back to the operating theatre, scrubbed up again and switched with Katsumata. He said nothing, but looked crestfallen – a clear message. I asked Mike to start ventilating the lungs and told the perfusionist to prepare to slowly ease off the machine. At this point Kirsty's heart needed to take over and support the circulation, otherwise she'd die on the operating table. We all stared at the traces on the screen, hoping to see her blood pressure rise. It briefly reached half normal, but then fell away rapidly as the pump was switched off.

'Shall we go back on?' asked Katsumata.

Watching the left ventricle flicker on the echo, the perfusionist questioned whether it was worth it. But 'She's had it, hasn't she?' was the real message from behind the drapes.

I wasn't yet ready to call it. Failure would mean death for the little girl and a life of torment for the parents.

'Let's go back on, give it another half hour.'

This in itself was problematic, as a long bypass time always lessened the chance of recovery.

Kirsty's parents were waiting in the children's ward – Archer had gone to warn them. When we called him back, Becky insisted on coming to the doors of the operating theatre complex with him. It's impossible to describe how a mother feels in these circumstances. All I knew was that the prospect of her holding Kirsty's emaciated and lifeless corpse was not far away. Should I tell her that the heart was too badly damaged, that the diagnosis should have been made months ago and that Kirsty had been let down by an overburdened system?

In her own words, these are the very powerful thoughts that Becky recorded in her diary at the time.

Dr Archer came to see us at hourly intervals. After around four hours I thought everything had gone well. Kirsty was to be taken off bypass and then transferred to intensive care. I popped down to the canteen to grab a sandwich but as I was making my way back one of the ward nurses was looking for me. She told me I needed to come back upstairs because Dr Archer was waiting to see us. I was really pleased and asked if the operation was done. Could we please see her? She looked very serious and said we must speak to Dr Archer. Though she was really kind and professional I guessed that something was wrong.

Back in the room a stern faced Dr Archer sat us down. He explained that despite the best efforts of the surgical team, Kirsty's heart would not separate from the bypass machine. The surgeons were still trying but the prospects looked bleak. We may lose her.

Then he had to leave. By now my head was spinning. I remember feeling sort of drunk. This wasn't the plan. If we waited patiently everything would be OK because that sort of thing only happens to other people.

Then Dr Archer came back. He told us he was so sorry. Every option had been exhausted. He would be arranging for us to go and hold her to say goodbye. I could not bear the thought that when I saw her again she would be cold. My baby was so soft and warm. Smelt delicious, hair like velvet, hot fuzzy cheeks. I just kept thinking that my heart would break if she was limp and cold. It sounds odd but it was such a strong feeling.

Obviously this was our darkest moment. The thought of Kirsty fighting for her life and nothing we could do. We might as well have been on the other side of the world. My frenzied brain went into overdrive. If she died they would put her on a cold slab in the mortuary. That hideously soulless place. If that happened I would stay with her until she was buried. I would fight anybody that tried to stop me. My baby girl would stay in my arms and God help anybody that tried to take her away from me.

Those thoughts remain as clear in my head as they were on that day because I never felt so strongly about anything. We had made a really close bond with other parents on the ward. All day they had been popping in asking for news, praying for Kirsty and sharing in our hopes.

When Dr Archer left our room no one else came in. I didn't blame them. There was a horrible feeling of sadness. Everyone was so involved in each other's journey and now nobody knew what to say.

On the very rare occasions that a child died on the operating table I always talked to the parents myself. It was something I dreaded, the very worst part of my job.

The sliding doors to the theatre complex opened automatically onto the hospital corridor. I was immediately confronted by eyes full of grief and desperation. I remember Becky saying, 'Please save my little girl.' I was poleaxed and said nothing. Archer looked desolate. He'd already done the difficult job. I turned back to the sombre theatre, put on a new mask and scrubbed up again.

Mike had finished yet another cigarette, and said, 'Things are no better. Can we turn the pump off?'

'No, I'm going to try one more thing. Turn the lungs off. Run the camera again.'

This was my last-gasp attempt. It was something that could only be justified by invoking the laws of physics and had never been done before in a child. The tension on the wall of Kirsty's scarred left ventricle was elevated because of the size of the cavity. From a recent conference I knew that a Brazilian surgeon had made a series of failing adult hearts smaller when a tropical infection, Chagas disease, had weakened the muscle. The operation had been attempted for other types of heart failure patients in North America but was quickly discredited and abandoned. In my view, this bold approach was Kirsty's last hope.

I was not going to risk stopping the heart again, so I took a glistening new scalpel and cut the beating left ventricle wide open from apex to base, just like unzipping a sleeping bag. I began in an area of scar, carefully avoiding the muscles that support the mitral valve, and the filleted heart immediately fibrillated in response to cutting. This was fine because there was no risk of it pumping air.

Frankly I was stunned by the unexpected appearance of the inner lining of the heart. It was covered in thick, white scar tissue. To reduce the diameter of the ventricle I cut away the tissue on either side of the incision until I reached bleeding muscle, removing one third of the circumference of the chamber. In an attempt to stop the mitral valve from leaking I sewed the central point of its two leaflets together, turning it from an oval to a double orifice structure resembling a pair of spectacles. Then I simply sewed the muscle edges together with a double row of stitches to close the heart. In the end, this much smaller heart looked like a quivering black banana. Not for a moment did I think it would ever start again – and nor did my colleagues. Most of them thought I was crazy.

Word of the bizarre operation in Theatre 5 soon spread. The curious gathered to watch and the camera kept on filming. We had to ensure that all air was removed from the heart, otherwise it could be ejected into the blood vessels of the brain and cause a stroke. After that, all that remained was to defibrillate and try to restore normal heart rhythm.

'We're done,' I announced. 'Try 20 joules.'

Zap! The heart stopped quivering, and for what seemed like an age there was no spontaneous electrical activity. I poked the muscle with forceps and it contracted in response. This time there was a flicker on the blood pressure trace. Miraculously, the black banana had ejected blood into the aorta.

Mike looked again at the echo. 'It certainly looks different. Shall we try using the pacemaker?'

I was already sewing the fine pacing wires into place. We arbitrarily set the pacing box rate at 100 beats per minute and switched it on. I told the perfusionist to cut back on the pump flow rate and leave blood in the heart to see if it would eject consistently. It did. What's more, the echo showed that the mitral valve no longer leaked. At this point I felt that we were in with a chance. Life really does depend upon physics and geometry.

It was now after midday. Kirsty had been on the bypass machine for more than three hours and we needed to get her off it. As if timed to perfection, her own heart rhythm suddenly broke through the pacing. Coordinated natural heart rhythm is much more efficient than electrical pacing, providing much better blood flow and pressure.

It was like switching on a light in the operating theatre. Gloom changed to elation. My own adrenaline kicked in and the fatigue suddenly lifted. We gave Kirsty an infusion of adrenaline to help her heart take over from the bypass machine. Finally I gave the instruction to 'come off slowly'. We still expected her blood pressure to fade, yet the curiously reconfigured little heart kept on pumping.

'Off bypass. I don't believe it,' said Mike.

I remained silent, but I looked over my mask at Katsumata. He knew that I'd had enough by now.

'Let me finish,' he said.

'Sure.'

I took a last, disbelieving glance at the little black banana pumping away, then turned to the echo screen, where the incomprehensible flashes of white, blue and yellow were also reassuring, like a blazing fire. We could see blood streaming through the new left coronary and a double jet entering the left ventricle through the mitral valve – a curiously reconfigured baby's heart that finally worked.

After the encounter at the theatre door both Archer and the parents believed that Kirsty had died. This created an unprecedented and awkward situation, but one that I was too knackered to deal with. I asked the anaesthetic nurse to bleep Dr Archer and tell him to come down yet again. She did this, then offered to get me a coffee.

Katsumata made certain that there was no bleeding, then meticulously closed the chest.

'Never been done before,' he said, looking over at me.

Soon afterwards Becky came down to the paediatric intensive care unit in shock. She put her hand on Kirsty's little foot and exclaimed, 'It's warm. It's never been warm.' When she started to cry I left. It had been a long day.

Dee, my endearing and eccentric secretary, drove me home to Bladon, some twenty minutes outside Oxford. I was restless, with a mixture of elation and fatigue. A huge, red, wintery sun was setting over Blenheim Palace. To wind down I set out to run around the lake with Max, our German shepherd dog. Through the ancient oaks we scattered rabbits and fortunate pheasants who'd survived the shooting season. The shadows lengthened. Hissing swans told Max to piss off. The sun slipped away as I staggered by the water's edge. I left the park at the Bladon gate, crossed the road and went into the grounds of St Martin's Church.

Winston Churchill is buried in the graveyard. Facing his tomb – invariably surrounded by dead flowers with their heads dipped in reverence – is a wooden seat donated by the Polish resistance from the Second World War. Hot, and gasping for breath, I sat down to talk with the great man, less than ten feet away in his box. Morbidly, I tried to visualise how his corpse looked now and contemplated how Kirsty at that moment could so easily have been stiff and cold in the hospital mortuary. But I'd followed his dictum – never surrender.

Max irreverently cocked his leg on a neighbouring tomb. Now I needed to sleep. I hoped that the phone would leave me in peace. It did. Kirsty survived.

We followed her carefully over the next ten years, using echocardiography to watch her heart develop. She was a perfect little girl, happy, outgoing and energetic, the only clue to her extraordinary internal metamorphosis being the faint stripe up the middle of her chest.

When we felt she was mature enough to discuss it we asked for her permission to carry out a magnetic resonance imaging scan to show us how the remodelled heart had developed. What we found was quite extraordinary. Apart from the double-orifice mitral valve her heart appeared normal, as did her new left coronary artery. Only a thin scar showed the position of the line of stitches up the heart. Remarkably, all other scar tissue had disappeared. The whole inner lining of her left ventricle had been pure white scar tissue – now all gone.

This provided some of the first evidence that an infant's own cardiac stem cells can regenerate heart muscle and actually remove fibrous tissue. Adult hearts cannot recover in the same way. But what if we could identify and culture stem cells that could do just this for an adult heart? Could it provide a solution for the hundreds of thousands of adult chronic heart failure patients with coronary artery disease? My grandfather was someone who might have benefited. We could inject the cells at the same time as their coronary bypass surgery, or get them in through a catheter within the heart. What cells would we use, where would we find them, and how could we preserve and transplant them? One day I'd find out.

Now aged eighteen, Kirsty is a vivacious and athletic teenager, but had she died we'd never have known about this exciting possibility for heart regeneration. Her case could potentially save countless other lives.

9

domino heart

I will give you a new heart and put a new spirit in you; I will remove from you your heart of stone and give you a heart of flesh.

Ezekiel 36:26

BARELY A DAY WENT BY without a visit to the paediatric intensive care unit to see the babies or small children I'd operated on, and to reassure their parents that one day soon things would get better. More often than not these visits provided a window onto other personal tragedies. Nothing could be worse than the black gangrenous limbs of infants with meningitis that required amputation. Or children injured on the roads with multiple injuries or brain death. Or the untold complications of cancer and chemotherapy. Why should kids get cancer? How can that be fair? Or hydrocephalus – babies with huge skulls full of fluid, their heads heavier than their bodies, unable to lift them from the floor? Fragile, miserable lives.

* * *

It was three weeks after our success with Julie. I'd been summoned by the paediatric cardiologists, who wanted to discuss an urgent case. Would I come straight away?

Several doctors were standing at the foot of a boy's bed, sifting through charts and test results. His desolate mother sat hunched beside him, her face strained with the anxiety of it all, holding his sweaty hand and staring at the heart monitor. He was propped up on pillows at a forty-five degree angle, eyes closed, chest heaving and making grunting sounds with every breath, coughing intermittently. I could see that he was deathly pale and limp, his eyes closed, his head tipped back, neck extended, struggling to breathe and quite obviously emaciated with that yellow tinge of terminal cancer. He was somewhere else in his mind.

So why did they need me? Maybe he had a tumour in his heart. Rare – certainly – but I'd operated on several heart tumours in children. Or maybe cancer had spread from a kidney or bone to the pericardium, causing fluid to compress his heart. I'd often been asked to make a window in the pericardium for such cases so that the fluid would drain into the chest cavity, where it did less harm.

Whatever the problem, it looked pretty desperate. I went unnoticed for a while – unusual for a cardiac surgeon – and just stood behind them and listened.

The boy's name was Stefan, and he was ten but looked younger. His mother had described him as 'not right' for some time – he couldn't keep up with his friends and wasn't concentrating at school. He'd even stopped playing football, because if he ran a few yards he was left gasping for breath.

Over the school holidays his parents had become increasingly concerned and he soon became really very unwell. The GP who'd listened to his chest said it sounded 'wet', then sent him straight up to the hospital for an X-ray. The news was bad. His lungs were wet because he had a huge heart with left ventricular failure and fluid on the lungs, what we call pulmonary oedema. This all came out of the blue as there was no previous medical history – no congenital heart disease, nothing to explain why he was now clearly dying.

The whole atmosphere was so tense that I needed to break the ice.

'Good morning,' I said. 'Can I help with anything?'

This was met by the usual response from Archer. 'Ah, Westaby, what kept you? Can I show you an echo?'

Stefan was concentration camp thin, with no fat on his chest wall, which told me that he'd been sick for months. His mum wasn't thin, so this wasn't deprivation. But the good news, if there was any, was that thin provides good echo pictures, and the problem was obvious when I looked at these. Both ventricles were dilated, worse on the left. The huge left ventricle barely moved and the mitral valve leaked. The two mitral leaflets were pulled apart, as what should have been a conical heart had turned into a sphere. A rugby ball heart, much like Kirsty's.

My thoughts went galloping on ahead. They were going to ask me to repair the valve and take the pressure off the lad's lungs. But surely this wasn't primarily a mitral valve problem. It was end-stage heart muscle disease, causing the valve to leak as a secondary issue. Any attempt at conventional cardiac surgery would finish him off, although I didn't state this out loud so as not to alarm his parents. Then I realised where this conversation was heading. It was about pumps.

By now everyone knew about Julie, who was still in the hospital but recovering well. We were beginning to get calls for help from all over the country. Viral myocarditis with chronic heart failure was Stefan's probable diagnosis, but because he'd been sick for months not days he was unlikely to recover as Julie had.

My immediate response was that he needed a new heart. Soon. Very soon. At the time only Great Ormond Street were doing children's transplants. I knew the surgeons well, as I used to work there. So let's get Stefan into their system and onto the urgent waiting list. Simple.

But not so simple, actually. Our people had already talked to the London transplant doctors and been told, sorry, but they were pressed for beds and already had several urgent patients waiting. And no, there was no chance at all of a transplant by order. Not in a child. Of course they'd get back to us when the situation improved, but in the meantime 'Do your best.'

Stefan was already receiving high doses of intravenous drugs to make his heart pump harder and diuretics to draw the water off his lungs. Without adequate blood pressure the kidneys won't work, and they were already struggling. He was on the verge of falling off the edge into the abyss. One of the paediatric cardiologists around his bed then asked the direct question: could I use another AB-180? With Julie we had set the precedent. If it really was myocarditis we might save him – and save his own heart. Or at least keep the lad going until Great Ormond Street could take him. It was the family's last hope.

I was conscious that the poor mother was listening to every word. The nurse had a hand on her shoulder, trying in vain to keep her calm. All eyes were on me. I went quiet and thought for a moment. Yes, we did have a second AB-180, but no, it wouldn't work. That inflow cannula was too big and too stiff to fit into a child's left atrium.

I shared this response with the gaggle of physicians. Their long, serious faces showed obvious disappointment, while the boy's mother started to cry. Archer had already suggested to her that the pump was the only outstanding option and that if the situation deteriorated further – as it was bound to – his course would be rapidly downhill, with an all-too-obvious terminus. So I'd just delivered the death sentence.

Stefan was an ordinary kid from a working-class family. He had his whole life ahead of him, and should have been in the school playground with his mates, not propped upright and petrified in an intensive care bed surrounded by white coats and long faces. He must have felt totally exhausted just lying there, the simple effort of breathing on its own leaving him worn out. Then that tedious cough with the tight feeling in his throat, like being strangled. He felt cold, yet his sheets were damp with perspiration. Strangers were sticking sharp needles into his arms and neck, sucking out his blood, shoving rubber pipes up his

private parts, things he should never have to contemplate
at that age. Seeing his mum and dad upset clearly disturbed
him, and he heard words that he didn't understand. Soon
he started to become light-headed, reality began to fade
and things slowly seemed to drift away into the far
distance. Morphine was taking the dread away.

His mum and dad slumped on either side of the bed,
leaning forward to be closer to him, both tense but
emotionally drained. They should have been at work, not
in the hospital – in fact they'd have rather been anywhere
else but here, with neither control nor influence as their
only son lay dying.

How could this have happened just out of the blue?
What did they do wrong? They'd now been told the brutal
facts – that the odds were poor – and they'd heard the
'transplant' word. Great Ormond Street had also been
mentioned. But nothing happens quickly. They could see
that Stefan was in shock and that his organs were failing.
Time was their enemy. Fear gripped their throats and lay
heavily on the chest to accompany their heart-rending,
gut-wrenching misery. Sentences became difficult, then
words, and soon it became impossible for them to speak
without an outpouring of emotion. But they tried not to
cry in front of the lad. Leave that for the end.

Archer was by this stage stressed and frustrated. He
knew the Great Ormond Street doctors well, and
although he understood that miracles are hard to come
by – other children were waiting in the same situation
with their own desperate parents – it would soon be too
late. He sifted through the blood tests. The potassium

was rising and so was the lactic acid, but he could neutralise this with sodium bicarbonate. Stefan would soon need kidney filtration. Archer did everything to prevent a catastrophic change in heart rhythm, which would certainly precipitate death. What else could he do in this hopeless situation?

The intensive care consultant waited in the background. Although he'd seen it all before, had looked after many children who'd died, he'd do his best. But what on earth would that be? Soon Stefan would need the ventilator as he was gasping for air, his breathing further depressed by the morphine. So the consultant hovered with anaesthetic drugs and breathing tube at the same time as he was waiting to do the ward round. He had nine other sick babies and children to worry about.

And then there was Stefan's nurse. Paediatric intensive care nurses are a special breed, as not everyone is cut out for the job of facing heart-rending anxiety and distress at work, day in, day out. A mature lady with children of her own, she liked to look after my heart surgery babies because they got better. She really didn't like to watch children die. Clearly she felt for Stefan's parents – the strain was beginning to show – and someone needed to do something drastic or it would be too late, her young patient's life ebbing away. She was the one who pressed Archer to find me.

The atmosphere around the bed was now as thick as fog, with a sense of impending doom. No one can snatch a donor heart out of thin air, especially for a child. Only a handful of children's heart transplants are done each year, so they'd looked to me for an alternative. But it wasn't going to happen.

I stared at the grieving parents and felt bloody useless. How would I feel in their position, if one of my kids were in the shit like this, struck down without the slightest warning? Their last hopes had just been dashed. Undoubtedly, having children of my own sensitised me to the plight of anxious families. By then my daughter Gemma was twenty and I had a son, Mark, at school in Oxford.

Viewing Stefan as if he were one of my children would be to empathise. Then he'd become a person, not just another patient. Some would suggest that empathy is the key to being a good doctor, the 'key to compassionate care', whatever that means. But if we really considered the enormity and sadness of every tragedy played out in this unit we'd all drown. That's why my intensive care colleague needed to get on with his ward round and not get sucked into the whirlpool that was Stefan's imminent demise.

Now I was rattled. At the time there was only one ventricular assist device suitable for children. It was called the Berlin Heart and had just recently been introduced by Professor Roland Hetzer at the Deutches Herzzentrum in Berlin. This was fortunate as he was a great friend – one of the benefits of scientific meetings – so I was going to call him and ask for a big favour, maybe tell him that Stefan was German. It certainly sounded that way. Added to all this, Roland was an Anglophile.

Fortunately he was right there in his office, so I got through at the first attempt. We exchanged the usual pleasantries then I came directly to the point.

'Roland, I need a Berlin Heart. The boy is ten but small for his age. There's a chance that the heart might recover but he won't last much longer. What will it cost me?' I knew that the money would have to come out of my charitable funds.

His response was as expected. 'Let's worry about that later. When do you need it?'

There was a brief pause. 'Could you get it to me by tomorrow morning, with one of your guys along to help?'

Roland was only too pleased to help.

The Lear Jet landed at Oxford airport at 8 am the following morning. In the interim I'd sent a message to our hospital chief executive with a copy to the medical director to announce my intentions. The insightful Nigel Crisp had moved on by then, and it was less than a month since I'd been threatened with the sack for saving Julie.

Archer did the honourable thing and went to see them both to try to persuade them it was our only option. The collective body of medical opinion had agreed that the boy could be dead by the end of the day, he informed them, and he'd exhausted all of the conventional channels. No one would help. If Westaby had a solution, he continued, they were morally obliged to let him get on with it. Action first, recriminations later. Oh, and by the way, had they visited Julie Mills yet on the ward? A world's first for Oxford, wasn't it? And if not, why not?

Archer was and remains a religious man. He held back from the 'resurrection from the dead' analogy and agreed with them that not only was Westaby not God, but that he might indeed even be an irritating son of a bitch. But wasn't it his job to save lives? That's all he was trying to do. So back off for now. Let the Germans come.

For my own part, I held fast to the concept that it was ethical to save lives whatever it took. I didn't need an anal ethics committee to cast doubt on that. And I didn't care about being sacked. I needed to work where I could fulfil my potential, pushing the envelope. If Oxford couldn't back it, I would go somewhere else!

The Berlin Heart consisted of an orange-sized blood receptacle divided into two parts – blood on one side, air on the other, forced inflation of the air chamber driving blood onwards through the valved tubing. Simple, but highly effective. The pump chamber sat outside of the body and could be exchanged if blood clots were detected within it. Inflow and outflow tubes connected the pump to both sides of the heart, and all the tubes passed out through the abdominal wall from the failing heart to external pumps. Then both ventricles were bypassed and rested, with guaranteed blood flow to the lungs and body. Just what the doctor ordered, I guess.

Now I needed to get Stefan down into the operating theatre. And not only that. The Lear Jet was waiting to take the German team home after the surgery and I was paying for it. Not exactly a black taxi cab with the meter running.

Stefan had managed to survive the night without being put on the ventilator. Now he was physically exhausted and very scared. At his age he had an appreciation of his predicament – he could read the long faces and his mother's tears – so there was an emotional separation in the anaesthetic room, the sort I prefer to avoid. Children's anaesthetists deal with this every day, but for me it's added pressure that I don't need, so I took the German team to change into theatre gear. This was embarrassing in itself – a scruffy room jammed full of grey lockers, brown wooden benches with paint peeling off, plaster coming adrift from the walls of the toilets, discarded theatre shoes, masks and clothing everywhere. What shoes would they wear? We searched around until we matched a couple of pairs, then went into the perfusionists' room to show them the kit.

Desiree was already there, ready to learn, and the surgical registrars were waiting with Katsumata. There was already an air of frenetic excitement, a sense of breaking new ground, something to tell their partners and kids about when they went home. Would it be on the news tonight? No. Maybe the local Oxford news? No. Would I be getting the sack? Quite probably. That would make the news. But we'd say nothing at this stage. Let's just get the boy better.

Stefan was a sorry sight as he was wheeled in on the operating table – so thin, almost pathetic. By this point I was sure that he wasn't suffering from viral myocarditis; it could only be severe chronic heart failure, with a heart muscle pathology that was unlikely to recover. Step one remained the same. First keep him alive, then take stock.

I ran the saw up his sternum and cranked the edges apart with the retractor. We opened the pericardium and tagged the edges to the skin, bringing the heart up towards us. Lots of straw-coloured fluid spilled out. I reckoned that about a quarter of his body weight was excess heart failure fluid, full of protein and salt, and now discarded by the sucker. I wondered if I was a fool to have become involved in this world of misery. There were easier jobs.

Now I had a good view of the struggling, dilated organ. His right atrium was tense and blue, ready to burst from the high pressure in his veins, and swelling his liver. His right ventricle was distended, and I looked carefully at the right coronary artery to rule out the possibility that he had the same problem as Kirsty. But he didn't, and in any case Archer would have spotted it if he had. There was no scarring in his huge left ventricle, just pale, fibrous-looking muscle that had given up. Nor was his heart swollen and inflamed like Julie's. I'd biopsy the muscle, then we'd see precisely what the problem was under the microscope.

The Germans were looking over the drapes at the top of the operating table. As two of Roland's elite transplant team, they'd seen many similar struggling hearts in Berlin. They used the generic term 'idiopathic dilated cardiomyopathy', a quite unusual condition in a child of ten.

It was already clear that Stefan would need both right- and left-sided support to keep him going. A left heart pump alone would deliver more blood to the body, but it would all come back through the veins to the right ventricle, which would then pack up because it couldn't cope. So right-sided support was essential. There would be four tubes coming out through the abdominal wall to two air-driven prosthetic ventricles that would passively fill, then forcefully eject blood, providing a similar volume and pumping rate to a normal child's heart. Good or what!

I sensed that his buggered organ would not tolerate manipulation as we'd provoke rhythm changes, making things worse before the pumps could be connected. I'd put him on cardiopulmonary bypass first to keep him safe. I then tried to lighten the atmosphere with a joke.

'I just deleted all the Germans from my phone.' Pause. 'Now it's Hans free!'

They didn't get it. Nor did Katsumata. We all pressed on in silence, making the four stab holes to bring the cannulas out of the chest, fixing one end to the heart and the other to the pump. Most importantly, we emptied the system of air. I attempted another joke.

'What's the difference between a hippo and a Zippo? One is very heavy, the other a little lighter!'

But once again, not a titter.

Everything was completed according to plan and the time for the switch-on had come. The prosthetic pumps were like normal ventricles but sat outside the body, where you could watch them work – lub dub, lub dub, lub dub. Energetic and effective. Stefan's own heart emptied like a deflated balloon and there was much better blood pressure now, with vigorous pulsation in the aorta and pulmonary artery. Lub dub, lub dub, lub dub. It was a ridiculously simple approach, but an outstanding result – a triumph of life over death. Although there was something more aesthetic and satisfying about pulsatile flow, with Stefan the pumps had to be external. At least continuous-flow devices were small enough to implant within the body.

Katsumata took care to ensure that there was no bleeding, squirting biological glue around the incisions to stop the tedious oozing. We needed to leave two drains in Stefan's chest to let the blood out, so there were now six tubes sticking out of his fragile little body. This meant he'd had multiple stab wounds, but they were all necessary. The usual thick, stainless steel wire stitches were used to draw the edges of the sternum together, pulled tight and twisted to cover all the hardware inside.

Stefan was then taken back to the paediatric intensive care unit, where his ventricular assist device was the first they'd ever encountered. It could have been intimidating for the nursing staff, but it wasn't. We told them not to focus on the tubes or the knobs on the console and that there was no need to change anything. All they had to do was take care of the boy, particularly as he'd be alarmed and agitated when he woke up.

We also emphasised that it was vital that Stefan didn't pull on the tubes, as they were his lifelines. When he awoke it would be best just to sit him up, take him off the ventilator and remove the tracheal tube – one less cause of discomfort. Then it would be possible to reason with him and be easier to keep him calm. His parents could sit with him and Desiree would be close by to help, even when she was off duty.

With that the Germans were gone and we were left on our own with the technology. No problem, as Stefan improved rapidly. His urine flowed again into the catheter bag and, as anticipated, he woke up in the early evening and had the tracheal tube taken out. He was pretty cross – even with his poor mother – but he was pink again, with rosy cheeks, warm legs and warm hands, which his parents held so tightly. He was just not keen on the aliens emerging from his belly, pulsating away in front of his nose – priceless, lifesaving technology, but pretty daunting for a kid.

As the days passed I was eager to know what the biopsy showed so we could work out the next step. The Berlin Heart would keep him alive for weeks or even months, but could his own heart recover? I suspected it wouldn't, so in the background we needed to plan for a transplant. Ever curious, I went to the pathology laboratory myself and asked to look at the processed specimens from Julie and Stefan, in the same spirit that I always went to the autopsies of patients of mine who died. The pathologists knew me well enough and they appreciated the clinical feedback.

Julie's heart muscle was densely infiltrated with a type of white blood cell called lymphocytes that respond to a viral infection. While viruses are too small to be visible under a light microscope, lymphocyte infiltration tells you that they're there. There were millions of them present, and the muscle was swollen and oedematous with the inflammatory process.

This was not so with Stefan, which came as a bit of a shock in a ten-year-old. Much of his muscle had been replaced with fibrous tissue, but not as a result of a lack of blood supply. There were no white blood cells at all. Stefan did have chronic idiopathic dilated cardiomyopathy – a long-term enlargement of the heart, cause unknown – and his condition would never improve simply with rest. He'd simply hit the buffers. Julie and Stefan had only one thing in common. We'd got there just in time. But the way forward was now clear – Stefan needed someone else's heart to go home with.

In those days – as now – neither an individual hospital nor a surgeon on their own could organise a heart transplant, not even if a suitably matched brain-dead donor was lying in the bed next to the patient needing the heart. There was a decision-making process to follow and an organisation to take on – the UK Transplant Service. They'd decided that, in order to make better use of scarce donor organs and to ensure equitable distribution, the 'urgent' category of patient should be abandoned. So, at the time, donor organs were offered to the transplant centres on a strict rotational basis. Many of those who received a donor heart were out in the community – not on a life support device like Stefan. We now know that these ambulatory patients gain little or no survival benefit from a transplant and many die from complications afterwards. Wasted organs, and one reason why I was driven to find an alternative. What's more, if a heart transplant took place involving an organ offered informally, then the transplant unit concerned had to report it to the UK Transplant Service, who would then place them right at the bottom of the waiting list.

I was growing increasingly concerned about finding a heart for Stefan and I had to get him into the Great Ormond Street system. I called the transplant surgeon Marc de Leval, whom I'd trained with and greatly respected, and who in turn was supportive of the fact that I'd developed a congenital heart service from nothing in

Oxford. Over the years I'd send him any complex case where I felt he could do a better job than me, as there's no place for pride or arrogance when operating on small people. I explained that we'd already tried to transfer Stefan before he'd taken a downward turn.

Marc knew all about this and was willing to help. He was also interested to see a Berlin Heart. Although Stefan was now stable, he was in such a parlous and unpredictable state that he could be added to the Great Ormond Street transplant list, just as if we'd managed to transfer him the previous week.

But there was a problem. Transferring him to London while he was on the Berlin Heart was going to prove unreasonably hazardous. When we asked the ambulance service to take him there, they couldn't guarantee sufficient electrical power to cover the transfer time, given the risk of getting stuck in traffic or the vehicle having a mechanical breakdown. So we'd need to make preparations with the Oxford transplant coordinator, confirming the blood group, arranging the tissue typing and looking for unusual antibodies in his blood. If we found a suitable donor heart we'd transplant him in Oxford, meaning that the medical director would almost certainly have a stroke.

The endgame came sooner than we expected, but we were ready for it. Stefan had improved day by day, still feeble physically but no longer in heart failure. The following weekend we received the transplant alert. Only thirty miles down the motorway, Harefield Hospital were preparing for a heart and lung transplant in a teenager with cystic fibrosis who was severely debilitated and dying from lung failure. She'd been on home oxygen for several years but was now bedbound, blue and gasping for air, with high pressure in the lung circulation and regularly coughing up blood. When she received a heart and lung transplant her own strong heart could be given to Stefan. This was the plan, a procedure like this being known as a domino heart, for obvious reasons. Domino transplants were rare then, non-existent now.

The cystic fibrosis patient was brought into Harefield while the organ retrieval team stood by. The logistics were nothing if not complex – the donor was many miles away and four separate surgical teams would be involved, for heart and lungs, liver and two kidneys. These would all be heading for different cities and were like vultures hovering above their prey, ready to consume the best bits of the body, albeit with the noblest intentions. They were all travelling at night and their journeys were not without danger, airborne transplant teams having been lost on occasion in bad weather.

When it was established that the heart was a tissue match for Stefan, and that he and the cystic fibrosis patient shared the same blood group, the process was scheduled to kick off during Saturday night. Could anything be better than that? We'd operate in Oxford on a quiet Sunday morning with minimal fuss.

Better still, the heart was not subject to the adverse physiological consequences of donor brain death. Head injury donors had often been subject to fluid restriction and diuretic therapy to reduce the pressure in their skull, and this, together with pituitary gland damage, often prompted the need for resuscitation with several litres of fluid. Many needed massive drug support to maintain adequate blood pressure, so as a result these compromised donor hearts often failed in the post-transplant period. I'd worked at Harefield for three years and knew the score.

The Great Ormond Street transplant coordinator would keep us abreast of the timing. The domino heart would be removed from the heart and lung recipient at around 7 am, and by the time it arrived in its plastic bags and cool box Stefan's chest would be re-opened and ready. He'd go onto the heart–lung machine to remove the Berlin Heart, and we'd chop out Stefan's useless organ, cannulas and all. My team would be in early and raring to go.

Ten was a difficult age to face something like this, but Stefan understood the situation, expressing relief at the prospect of having the aliens out, and then resignation. And trepidation. He hated having four tubes the size of hose pipes sticking out of his belly, blue blood streaming through one pair, then bright red in the other, as well as noisy, pulsating discs in front of his nose. We'd originally told him that he might need to spend months with all this equipment on, so the early transplant was a happy release.

But we didn't tell him the risk of not getting through, which in those days was 15 to 20 per cent following failure of the donor heart, infection or rejection. But this domino heart – from a live person with a normal brain – was particularly strong. And it was a reasonable match on the tissue typing. No alarm bells. We just needed to get on with it. Stefan's parents had been sitting with him since

6 am, and had spent most of the night awake, getting more and more anxious despite their underlying hope. As the tension mounted for them, they transmitted this anxiety to their son.

I took Marc to meet them. By now they were with Stefan in the anaesthetic room, where there was not a lot of space with the equipment and everything else. Marc's eyes kept flicking back and forth to the Berlin Heart, at this stage the only ventricular support system suitable for small children. Great Ormond Street needed to get one so that lives were not wasted.

Katsumata appeared at the door with news – the domino heart had left Harefield. With the light Sunday morning traffic it would reach Oxford in thirty minutes, so it was time for Stefan to be put to sleep. The moment of parting had now come, acutely distressing for the parents and – briefly – for Stefan. Kate, the anaesthetist, was poised and ready. Anaesthetic injected into the drip and soon his mental anguish was gone. A nod from Louise, the anaesthetic nurse, and the parents shuffled out of the door. Huddled together, their distress would continue for a while. As if they hadn't had enough already.

After that things happened quickly. My scrub nurses Linda and Pauline painted his chest with pink chlorhexidine antiseptic solution and dried it because the liquid is inflammable. They then covered him in sterile green drapes. Marc, Katsu and I scrubbed, gowned and gloved. The clock was ticking.

We unpicked Stefan's skin stitches, cut through the sternal wires and carefully inserted the retractor amid the many pipes. As with all sternal re-entries there was blood clot and fibrin sticking to the heart and tubes, so we peeled all of this off and sucked it away, then washed the heart and pericardium with warm saline solution. Everything needs to be clean – a tidy house for its new occupant, not a rubbish tip – and we had to find space for the cardiopulmonary bypass tubing. Once this was attached, we could switch off the Berlin Heart, cut through the tubes close to the heart and remove them from the surgical field.

But we weren't about to do that without the donor heart being in the room. There could still be a disaster en route – a road-traffic accident, a puncture, anything. Or someone might drop the heart on the theatre floor. This had happened before, with Christiaan Barnard in Cape Town. His brother Marius dropped it while taking it from the donor in one operating theatre to its recipient next door. Oops!

At 9.15 am the heart arrived in its box, surrounded by bags of ice. We set it down on its own table and unpacked it carefully one bag after another, finally allowing it to rest in a stainless steel dish. It sat there in its salt solution at 4°C, cold and floppy, like a sheep's heart on the butcher's slab. But we knew how to revive it, and had full confidence that it would start again and do its job. So I told Brian to switch off the Berlin Heart and go onto cardiopulmonary bypass.

Stefan's own heart emptied out for the very last time, then flopped down completely useless in the back of the pericardium. Marc started to trim the donor heart while

I chopped through the four plastic cannulas. Katsumata pulled them out of his body and threw them away. It was now time to chop out Stefan's sad heart ready for the new one. Out it came, leaving the empty pericardium – a curious sight. No heart. It must have been really scary when Barnard did it for the first time, like peering beneath the bonnet of a car and finding no engine.

The donor heart was implanted in a strict sequence, and it was essential to align it correctly without distortion. This may sound obvious, but donor hearts are slippery, wet and not easy to hold in position.

It helps to have a clear, three-dimensional vision of the finished product. I'm lucky in this respect as I inherited co-dominant cerebral hemispheres, meaning that I use both sides of my brain's motor cortex. I can operate with either hand. I'm a right-handed writer but a left-handed batsman, and I preferentially kick a ball with my left foot. Co-dominance helps with many things but especially surgery, being more important than the ability to study and pass exams.

But a heart transplant is quite simple. Take deep, full-thickness bites of the donor and recipient atrial tissue and keep stitching very carefully so there are no leaks. With the atria and aorta sewn together the aortic crossclamp can be released. This marks the end of the 'ischemic' period – the critical time that affects survival, during which the heart doesn't have coronary blood flow after removal from the donor. We know that the hearts that do best come from young donors with a short ischemic time and blood group compatibility. But that doesn't help

much. The patients have to take what they can get –
they're lucky to receive a heart at all. That's why even
'marginal' donors are accepted these days: the over-sixties,
smokers, even those with some types of cancer.

But it was all looking good for Stefan. Blood coursed
through the coronary arteries and brought the heart
muscle back to life, turned it from flaccid and pale
brown to almost purple, stiffened and fibrillating. As it
began the recovery process we stitched the last join
between the severed pulmonary arteries, then made
further efforts to remove air. Air in the brain wouldn't
help him.

At Marc's suggestion we rested Stefan's magnificent
new heart for an hour on the bypass machine. This
precious organ might easily have gone into the bin with
the diseased lungs. Its continuing life is one of the wonders
of modern medicine. It defibrillated spontaneously and
started to eject blood, gathering strength with time, then
separated easily from the bypass machine.

There were now two main risks. First was the rejection
of the donor heart, should the immunosuppression prove
inadequate. To balance this was the second, that excessive
immunosuppression might lead to serious, even lethal,
infection. So when Stefan recovered he needed to go to the
experts at the transplant centre at Great Ormond Street.
We'd done our bit by keeping him alive. Marc would let
us know as soon as a bed became available.

Archer and the paediatric intensive care unit helped us
to take care of Stefan over the next week, then he was
transferred to London. We kept in touch and followed his

progress. After a few transient rejection episodes that soon got better, he enjoyed a virtually uncomplicated recovery from what was an almost impossibly difficult starting point. We're still following him nearly twenty years later. He now has his own little family and is reaping the benefits of an ideal donor heart transplanted quickly, thanks to my friends in Berlin and Great Ormond Street.

Those few balmy weeks in summer were epic pioneering days. We'd achieved the UK's first bridge to recovery in viral myocarditis, then the first bridge to transplant in a child. These were dire emergencies undertaken on the hoof and worked through in the dead of night with my dedicated team of overseas colleagues. Great Ormond Street adopted the Berlin Heart for their heart transplant programme, initially with charitable funds. Then it became the only approved system to support babies and children with severe heart failure in the United States. It still is. Needless to say, we never got to use it again in Oxford. Children with heart failure either reached Great Ormond Street in time or they died. Julie and Stefan emptied out my own research funds. But what price can you place on two young lives?

10

life on a battery

We will now discuss in a little more detail the struggle for existence.

Charles Darwin,
On the Origin of Species

IT WAS A WARM SUMMER'S MORNING in the first week of June at the turn of the millennium. At 11 am there was a tentative, almost apologetic knock on my office door. And there stood Peter, his large frame filling the doorway. He leaned on a stick, swaying unsteadily, and he was sweating profusely, his head bowed, his lips and nose blue, panting for breath. Out of pride he refused to be pushed through the door in his wheelchair. Only weeks before he'd received the last rites, but such details still mattered to this man. Desperately trying to disguise his distress he slowly lifted his head and stared straight ahead through the doorway. He couldn't see me yet but – like Stefan – he reminded me of a concentration camp victim, a dead man walking, all hope abandoned.

My secretary Dee was visibly shaken by Peter's distress, so I broke the silence.

'You must be Peter. Please come in and sit down.'

Hidden behind the stooped frame was Peter's foster son, who parked the wheelchair in the corridor. I tried to make them comfortable with a little joke.

'Did you pay for that parking space? This is the NHS, you know!'

They didn't get it.

Peter shuffled slowly through to my room, and began staring at my certificates, awards and other surgical paraphernalia on the walls. He was checking me out. A religious man, he worked as a counsellor for the terminally ill with AIDS. But life had come full circle and he faced death. His existence had become that of an intelligent mind attached to a body rendered useless by heart failure. He was expecting the end to come soon, the sooner the better. I gestured to the armchair. He set the stick aside and sat down with a grunt.

Now I was checking him out. He was breathless on the slightest exertion, his belly bulged with an engorged liver and fluid, and I could see that his legs were swollen and purple. He wore oversized sandals, with socks stretched over massively swollen feet, and there were stained dressings on leg ulcers that the socks failed to cover. I didn't need to examine him. This was gross end-stage heart failure. I was amazed that he'd made the effort to leave home as he could die at any moment.

Some months before Peter's visit a colleague and I had written an open letter to members of the British Cardiac

Society (as it was then) to announce that we were ready to test a revolutionary new type of artificial heart – the Jarvik 2000. We needed to recruit terminally ill heart-failure patients who were not eligible for cardiac transplantation. Peter fitted the bill perfectly.

I'd already read his medical notes from the cardiologist. Peter had first been diagnosed in March 2005 with dilated cardiomyopathy that had been triggered by a viral illness affecting his heart muscle. He'd had a bout of influenza, which turned to myocarditis, but initially recovered. Or so it seemed. Now he had an enlarged, flabby heart, an irregular heart rhythm and a leaking mitral valve. Such patients usually die within two years of diagnosis, and Peter was well beyond this. He'd been admitted to hospital on many occasions, gasping for breath and coughing up fluid, and without rapid treatment with diuretic drugs this 'water on the lungs' would be his terminal event.

On each occasion his drug treatment had been esca-lated, with modest yet short-lived relief. Now he'd reached maximal levels of all useful drugs and his single kidney was failing. Months earlier his cardiologist had asked the surgeons at a London hospital whether they'd repair his leaking mitral valve, raising Peter's hopes. That was until his outpatient visit, when the surgeon was thoroughly dismissive, saying it was quite impossible, being far too late and far too risky.

The hospital correspondence described him as grossly fluid overloaded, breathless and exhausted on minimal exertion, unable to lie flat and only able to sleep propped

up on pillows or sitting in an armchair. This was exactly
as I remembered my poor grandfather.

Back in my office Peter was still sweating, as he tried to
regain enough breath to speak. I remember thinking that
this man would be lucky to survive a haircut and was
amazed they were really expecting me to operate on him.
But that's what mechanical hearts are for. This was
precisely the intolerable existence they were meant to
improve, the symptoms they were designed to relieve and
the life they aimed to prolong. By now Dee was less flustered and had brought tea. Peter thanked her. Now we
could talk.

I thanked Peter and his son for making the huge effort
to come, then asked him the circumstances of his referral.
He'd been working as a psychologist at London's
Middlesex Hospital and ironically was writing a book
called *Healthy Dying*. Just a few days earlier, he'd struggled
to a meeting with his co-author, Dr Robert George, a palliative medicine consultant at University College Hospital.

Peter wanted to say a last goodbye but was in so much
discomfort that Rob went to find a cardiologist to see if
anything could be done. While waiting for his colleague
to finish with a patient he glanced at the cardiologists'
notice board and saw a cutting about the heart pump
project in Oxford. He recognised the name of the surgeon,
Steve Westaby, as he'd known me as a junior doctor. Both
he and the cardiologist then wondered whether I could
help Peter.

Coming directly to the point, I suggested that we could
both help each other. I'd just been given the opportunity

to do something that had never been done before, something that had the potential to help hundreds of thousands
of patients worldwide if it worked. I bluntly told him that
I needed a guinea pig and that he'd be perfect.

I took the Jarvik 2000 out of a desk drawer to show
them. The titanium turbine was the size of my thumb or
a C size battery, and I explained that the pump would fit
inside his own failing heart, implanted at what used to be
the pointed apex. His left ventricle was now so large that
there was plenty of room, so we'd sew a restraining cuff
onto the muscle that would hold the pump in place, then
punch a hole through the heart wall and slide the pump
in. The high-speed turbine would empty his struggling
heart through the graft and into his aorta, the major
blood vessel to his body.

I showed him how the torpedo-shaped impeller spun
within the tube. It went at an unbelievably fast rate –
between 10,000 and 12,000 rpm – pumping five litres or
more of blood per minute, as much as a normal heart but
with continuous flow. It didn't fill and empty to eject
blood as the normal heart did, as there would be no pulse.
The only potential problem was that the right side of his
heart would have to cope with the boosted circulation,
but if the right ventricle coped well enough this manmade
pump could be as good as a transplant. If it didn't, he'd
die.

Peter winced at the word 'transplant'. No one should
underestimate the profound psychological trauma of
being turned down for a transplant, a patient's last hope
when life appears to be approaching its end. He was bitter,

as he'd been through the selection process twice. The first time he'd been told that he wasn't sick enough for a heart transplant. The second, when he was fifty-eight, that he was too sick.

I tried to put this into context for him. Assessment for a heart transplant is a brutal process. To describe transplantation as the 'gold-standard' treatment for heart failure is equivalent to claiming that a lottery win is the best way to make money. First of all, heart transplantation is ageist. In the 1990s patients older than sixty were not even considered. There were around 12,000 severe heart failure patients younger than sixty-five in the UK but fewer than 150 transplants. Clearly it was the transplant physician's responsibility to select patients who'd accrue most benefit, and there were very few of these.

What I wanted to do was to help patients in Peter's position – the desperately ill who'd never get the transplant opportunity, and those of all ages who were abandoned to 'palliative care', narcotic drugs blunting the misery of their unpleasant, lingering death. Peter had refused these drugs. He informed me that he was all too familiar with death, having comforted more than a hundred patients during the last days of their life, 'telling them the things they need to do, can do, the stages it will take, things like that'. It wasn't the time to compare body counts. I'd already dispatched more than three times that number to the afterlife.

Now rested, he had the measure of me and was more animated, an extraordinary character beginning to shine through his morbid preamble. His smile penetrated through his grey facies and purple nose, and I warmed to the man. So traumatised was he by repeated rejection that he had absolutely no expectation from our meeting. Quite to the contrary. He expected to be turned away.

I had serious doubts whether he'd survive a general anaesthetic, but if we took him on no one could claim that we'd picked an easy patient or one who didn't need the pump. Both my own hospital's ethics committee and the Medical Devices Agency had requested independent verification that the first patient to be given the Jarvik 2000 should be terminally ill, with a very short life expectancy, and Peter wouldn't fail on either of these criteria. So the decision rested with me. Impulsively I told him that it would be a great privilege if he'd allow us to help him and that if he wanted the first pump it was his. There followed a look of astonishment that quickly turned into a broad grin. This was his lottery win.

He asked about the odds. I said around 50/50, but knew that to be optimistic. Like many patients his main worry was that he'd be left brain damaged and worse off than to start with. I reassured him that if the operation didn't succeed he'd definitely die. A strange way to reassure someone, perhaps, but he was quite taken by the concept that failure spelt death. Life at present was unbearable, but as he was a Catholic, like most in his position he wouldn't contemplate suicide for his family's sake. Surgery was an option for euthanasia without the moral debate.

I asked about his wife. Why had she not come with him? Diane was a teacher and couldn't come away at short notice. Together they'd founded the National Association for the Childless, written the book *Coping with Childlessness* and had looked after eleven foster children. As a younger man he'd played rugby, something we had in common. I felt that he was a good person and would make the best out of his extra life.

I showed him the equipment and asked whether he could cope with life on a battery. He'd have to carry the controller and batteries – fitted into a shoulder bag – at all times. There was an alarm that sounded when the batteries were low or disconnected, and he'd need to change them twice a day. Overnight, he'd plug into the electricity mains supply at home. Very futuristic.

Now the next surprise. Dr Jarvik and I had worked out a revolutionary new method to bring electrical power into the body. The big issue with electrical power lines that emerged through the abdominal wall was their propensity for infection, as constant movement of the cable through fat and skin allowed bacteria to enter, and sometimes even the pump became infected. Seventy per cent of patients were eventually troubled by this and many needed further surgery. Instead we planned to screw a metal plug into Peter's skull. The scalp skin is virtually fat-free and has a generous blood supply. The plug would be rigidly fixed into bone, and we believed that this combination would minimise the risk of power line infection.

So Peter would have an electric plug in his head carrying electricity to the pump via a cable passing through his neck and chest. Magic! I'd be the real Dr Frankenstein.

Peter laughed. His mood was changing. I explained that he'd have a large, painful incision around the left side of his chest to implant the pump. Not so funny. There would be additional small incisions in his neck and scalp to fix the electrical system. Peter asked whether this had ever been done before. It hadn't.

'So, is it going to work?' he asked.

'Yes. I've done it in sheep.'

He laughed again, then enquired whether he'd hear or feel the pump in his heart.

'Well, the sheep never complained.'

It occurred to me that I should warn him that he wouldn't have a pulse. The impeller – the moving part of the pump, which spins at high speed – would simply push blood through his body continuously, like water through a pipe and very different from the pulsatile ejection of the biological heart. So his nurses and doctors would never feel a pulse or be able to measure his blood pressure? Yes. Life would be different but arguably preferable to the inevitable alternative. He was the pioneer in this respect.

Then another intuitive question – if he lost consciousness away from the hospital how would anyone know whether he was alive or dead? This was moving away from my comfort zone, so I sidestepped with a speculative response. But he was right to ask. Months later over the winter another pump patient fell and hit his head at home. He was found some time later, unconscious, cold and pulseless. The ambulance crew took him directly to the mortuary.

Now Peter's final question – was I nervous about attempting the operation, a procedure that was pure science fiction and quite likely to snuff him out?

'Absolutely not,' I replied. 'Not if you want me to do it. I'm not the nervous type. Doesn't suit the job.'

These words prompted a direct response.

'Let's go for it then.'

I told him that he should take just a little time to discuss it with his family and friends.

There was one more thing. I needed to see echo pictures of his heart myself. We pushed him round to the cardiology department and helped him up onto the couch. He was breathless again, and we could soon see why. His huge left ventricle was massively distended and barely moving. The mitral valve was held open by the stretched heart wall but this wouldn't matter with the pump in, as long as the aortic valve was not leaking – and it wasn't. The pump simply sucks blood through. His right ventricle seemed to be working well enough, and all in all his anatomy looked good for the surgery. I just needed to stop focusing on the risks. Failure was not an option, as death in the first patient would destroy the programme.

Peter got himself down from the couch and insisted on walking to the door. It would be wrong to say that he had a spring in his step, but he left with something much more important – hope. Hope for the first time since staggering away from the transplant assessment in despair. Now we simply had to get on with it.

Peter's wife Diane and some of the foster children took part in emotional discussions. Should Peter hang on for the short time he had left or risk dying during surgery for the chance of a better life? Diane told her husband that she couldn't make the decision for him or tell him what to do, but that whatever he decided she'd give him all her support.

Two days after our meeting Peter confirmed that he'd consent to the operation. Now I needed to ask Philip Poole-Wilson, Europe's leading heart failure cardiologist, to confirm Peter's poor prognosis. He could come to Oxford late on the evening of 19 June. Confident as to what he'd say, I reckoned that we'd proceed with surgery on the 20th.

I had to coordinate the team from Houston and New York. Bud Frazier, who'd undertaken the animal work at the Texas Heart Institute and had implanted far more mechanical hearts than any other surgeon, would be an important member of the surgical team. Dr Jarvik himself would bring the device over from New York, and we'd admit Peter to the hospital two days before the operation. We needed to optimise his heart failure treatment and teach him how to manage the controller and batteries. It was equally important that other members of the team should get to know him.

On the afternoon before surgery we bought Peter to the cardiac intensive care unit. Sister Desiree shaved the left side of his head in preparation for the skull pedestal incision. Dave Pigott, the anaesthetist, inserted a cannula into the artery in Peter's wrist, followed by a large-bore venous cannula into the internal jugular vein on the right side of his neck. He then floated a balloon catheter through the

veins across the right side of the heart and into the pulmonary artery.

I brought Jarvik and Bud to visit Peter in the early
evening. The conversation was animated and uplifting for
a man who faced 50/50 odds in less than twelve hours.
For the first time in months he talked about his future –
what he could do to support our programme if he survived
and where he would go for his first holiday in years.
Positive stuff, which helped all of us. Now we needed the
professor to come.

Philip arrived at 10.30 pm. He talked at length with
Peter, looked through the data, then emerged just after
midnight. He wished us luck. Adrian Banning, Peter's
cardiologist in Oxford, likened his predicament to that of
a man on a diving board who was about to jump but had
no certainty that there was water in the pool. According
to Adrian:

> Houghton was functionally dead. All he had left was
> a mind full of frustration. Heart failure has a worse
> prognosis than any type of cancer. Once you have
> fallen off the threshold for the transplant waiting list,
> conventional medicine has little to offer. Every
> cardiologist has clinics full of these people, unable to
> work, just hanging on, waiting to die.

We all assembled in the anaesthetic room of Theatre 5 at
7.30 am. As usual, Bud arrived in his Stetson and cowboy
boots – normal for Texas, less so in Oxford. I asked Peter
whether he had any reservations or last thoughts. He

replied that he'd be in a better place after the surgery, one way or another. I glibly told him that he'd be fine, something that every patient should hear before their anaesthetic.

Once he was asleep, we positioned him left side up on the operating table with the side of his head and neck exposed, and I marked the anticipated site of the surgical incisions with indelible black marker pen. We'd bring the power cable out of the apex of his chest, through his neck and to the left side of his head. Andrew Freeland, my cochlear implant specialist colleague, would screw the pedestal to the skull, while we'd expose the pericardium and aorta through the left side of his chest. This needed a large incision between the ribs.

With a degree of trepidation I exposed the leg artery and vein in Peter's groin to attach him to the heart–lung machine, then performed the chest incision through fat and wasted muscle. The metal retractor cranked open the ribs, bringing the lung and pericardium into view. Behind the lung was the aorta. Through a separate wound on the shoulder we passed the black insulated power cable up into his neck, then through the neck and out behind his left ear. This was difficult, as there was some important clockwork in close proximity to large arteries and veins, not to mention the vital nerves.

On the end of the electric cable was a miniature three-pin plug. Literally. This plug was inserted through a titanium pedestal that had six screw holes to fix it rigidly to the outer table of Peter's skull. Andrew made a C-shaped incision behind the ear and scraped the fibrous surface off the bone. Then a power drill was used to create the screw holes in his skull. The plug was screwed securely onto the skull with dry bone dust incorporated to promote healing around the titanium. We were making it all up as we went along.

All that remained was to punch a hole into the centre of the skin flap through which the plug would protrude, so that we could plug in the external power line leading to the batteries and controller. The head and neck incisions were then closed and we were ready to implant the pump itself.

I opened the sac around Peter's heart. It was a sorry sight. The huge, quivering left ventricle was more fibrous

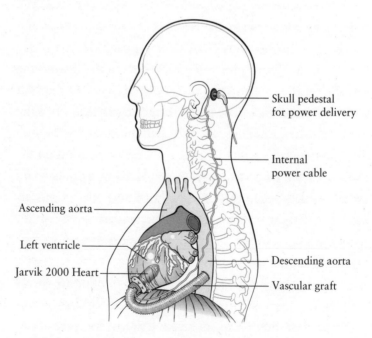

The Jarvik 2000 Heart implanted into the left ventricle – the operation performed on Peter Houghton

tissue than muscle. It barely moved, and by now, an hour into the surgery, Peter's blood pressure was uncomfortably low, lactic acid was building up in his blood and we had to start the bypass machine to support the circulation. Bud held the titanium pump, while I pulled the lung forward to expose the aorta. We needed to sew the graft end onto the aorta before the pump could be implanted into the heart. And the graft had to be just the right length – not so long that it could kink, but worse still if it was too short. What's more, the stitching had to be pristine to avoid bleeding.

Now we were ready for the big event. We started to stitch the restraining cuff onto the rounded apex of the heart, which resembled nothing less than a rotten melon. Never again would Peter's heart bear sole responsibility for his circulation. From now on his life was reliant upon technology.

All that there was left to do now was to core out the plug of heart muscle through the centre of the cuff then insert the pump, just like coring an apple and sliding in a torch battery. This was Peter's life raft. We were about to create a pulseless human being, and so far it was going well. I cut a cross in the muscle encircled by the cuff, and with a coring tool we punched out the hole and slid the pump through. It was in. The plan had worked – so far.

Desiree was holding the controller and batteries, waiting for the instruction to switch it on. Once I was satisfied that there was no more air in the pump or vascular graft we turned up the pump speed to 10,000 rpm, the flow probe showing that it was now pumping four and a half litres per minute. We then cut back on the heart–lung machine flow to let the combination of the device and Peter's own heart take over, slowly shifting support from one system to the other. I finally told Brian to 'Come off.' The whole process up to this point had taken two hours.

All of our attention was now fixed on the monitor screen. The arterial pressure trace was an absolutely flat line measuring no more than two-thirds of normal blood pressure, and pressure in the veins was also less than normal. While this suggested that the right ventricle was managing well, it was still too low. Peter's circulation needed to be well filled, otherwise the powerful turbine could suck the left ventricle empty, causing an obstruction. We aimed for a balance whereby the pump did most of the work but Peter's left ventricle continued to eject some blood.

Now we needed to adapt our treatment strategy to a completely new pulseless physiology – flatline physiology. We'd looked after a lot of sheep, so we knew just how to handle it.

The remaining and most troublesome issue was to stop the bleeding. Every cut surface and every needle hole was oozing blood because Peter's distended liver had not been making clotting factors, an issue common to most patients who needed artificial hearts. So we gave donor clotting factors and the sticky cells called platelets that plug needle holes, leaving the registrars to close his chest.

Outside the theatre we reviewed the power consumption, which was seven watts. The pump flow oscillated between three and a half and seven and a half litres per minute, the rate depending upon the pump rotor speed and Peter's own blood pressure, now providing resistance to pump flow. This was counter-intuitive physiology: when Peter's blood pressure increased, flow decreased substantially. If flow to the body and brain was insufficient, lactic acid would accumulate in the blood and the kidneys would stop producing urine. But right now things were all right. The pump was doing its job.

With his chest closed, the drapes were removed and Peter was placed onto the trolley for transfer to the intensive care unit. We had an elite nursing team who were carefully prepared and knew what to expect. He was hooked up to the monitor and an audience gathered to see the flatline patient, the first to be fitted with a revolutionary new type of artificial heart on a permanent basis. We left the nursing team in charge, with orders to call us if anything went awry.

I was overstimulated after one of the most exhilarating operations I'd ever performed and could barely sleep. So, when the sun rose at 4.30 am, I visited Peter in his room. Listening to his heart with a stethoscope there was now no lub dub, lub dub, lub dub, just the characteristic continuous whirr of the pump rotor. His single functioning kidney had stopped producing urine, but we expected that. What worried me most was that blood transfusion is bad for the lungs and he'd already had thirty units of the stuff. With blood now flowing backwards up the descending aorta to the brain I wondered how long it would take him to wake up. Only time would tell.

Peter remained stable for the next thirty-six hours and began to regain consciousness. As soon as he was sufficiently awake to breathe, cough and understand instructions, we propped up his large frame and took out the breathing tube.

The first thing he said when he saw me was 'You bastard.' A thoracotomy between the ribs is very painful, and he had other incisions in his head, neck and groin. But it was said with a smile on his face, with a sense of humour. He was glad to be alive. We talked for a while about how the operation had gone. I joked that despite his Christian faith he was now Frankenstein's monster, powered by a bolt in his head, the thing that was currently giving him a headache. But he was extremely motivated to get well and make the most of his new life.

Within the first week his kidney function improved and we no longer needed to dialyse him. He worked hard to get out of bed and regain mobility with the help of the physiotherapists. Although the pump immediately restores blood flow to normal levels, it still takes months to reverse the debilitating wasting effects of chronic heart failure. It's just the same as with a transplant. But it was already a source of wonder and relief that his breathlessness had gone, and that there was no more back pressure from the failing left side of his heart on his lungs. He began to lose the litres of fluid that had chronically accumulated in his tissues, his leg ulcers began to heal, and his face and nose were now pink, not blue.

Remarkably Peter left hospital just eleven days after his operation, taken home to Birmingham by his family. Such a swift departure could have never happened in the States. Before he left, Peter appeared in front of the press, with numerous photographers waiting at the hospital entrance. He was in his element and enjoyed himself immensely. Our Anglo-American team had achieved a world's first, but Peter was the star – the bionic pulseless man. He described himself as the model cyborg.

Peter's ability to exercise increased progressively. Within weeks his belly began to shrink as the fluid around his guts dissipated, then his huge legs became slim again. At an outpatient appointment five months after the operation in November even his heart rhythm was back to normal.

He was really chatty, telling me that events since June had taken him from the position of a refugee, forced to pack up and abandon all the trappings of life, to that of a man given uncertain leave to remain. All the while his engaging personality shone through. He'd shifted from inexorable fear and bewilderment to undisguised pleasure at avoiding death, and was now fitter and healthier than he'd been in years. He reflected:

It really annoys me when people say I was brave; I wasn't brave at all. I was swapping a certain slow death for the risk of a quick one, or the chance of a complete recovery. When I first left hospital I didn't dare plan ahead. I was literally living my life from one day to the next. Now I'm thinking about what to do with my time, and contacting all my friends to tell them I'm not dead.

Out and about in Birmingham, Peter was a curiosity. It took time for the hair to grow back on the side of his head, and at first his plug and black power cable were obvious to passers-by. Children would come up to him and ask why he had a bolt in his head. Was he a robot? Peter was more than happy to stop and explain things to them. He had a particularly happy Christmas that he never expected to see.

Out shopping one day during the January sales he felt a sharp and painful jerk at his head. A would-be thief had snatched at the shoulder bag containing his controller and batteries, thinking it contained a camera. His skull pedestal plug was torn off and his pump stopped. The teenage mugger attempted to flee with the bag in his hand but the power-failure alarm sounded noisily. Sensing a trap, the lad discarded it and ran off. Some shoppers helped to retrieve Peter's power cable and he fumbled to get the plug reattached to his head as fast as possible. An old lady sorted it out for him but was bewildered by what she'd achieved. Reconnected, the pump whirred undeterred.

'I did feel faint,' Peter recalled, 'but I think it was more the shock than anything else. The wrench to my head really hurt for several days.'

He spent the first year getting as physically well as possible, then the second finding some meaningful purpose to make his 'extra life' worthwhile. This second chance would constitute more than 10 per cent of his overall life span. It was vital for him to have a purpose for his existence beyond being an exhibit, so he worked tirelessly to raise money and draw attention to our efforts, desperately keen that others should benefit from the same opportunity. Soon he became an integral part of our team, counselling other potential assist device patients and their families.

Peter was never the most compliant of patients. He suffered from nosebleeds and would reduce his anticoagulant dose to cope with these. And then there was the price to pay for his reprieve: he needed to change his batteries every eight hours, recharging the spent set, and he had to carry his equipment everywhere as a matter of routine. Sometimes he'd forget to change the battery for a fully charged one before going out. On one occasion he was in the middle of having a tooth filled when the battery near-expiry alarm went off and the dentist had to drive him home rapidly.

He was a prolific writer and published his own book, *Death, Dying and Not Dying*. He gained considerable satisfaction when his charity was able to contribute to pump implants for other patients, then greatly enjoyed the camaraderie of his new bionic colleagues, most of whom resumed active and sometimes adventurous lives.

In the back of his mind he always hoped for his own heart to recover sufficiently to enable the removal of his hardware. Although this happened up to a point, we resisted the temptation. This was fortunate since his heart deteriorated again, and for the last three years of his life he was wholly reliant on the device. Ironically, he was offered a heart transplant but pointedly refused to discuss it.

In his sixth and seventh extra years he was troubled by ageing problems that at one time he'd never expected to encounter. He developed rheumatoid arthritis in the hands, which hindered his writing, and his prostate enlarged to the point that it required surgery. We arranged this for him in Oxford, since no other hospital would operate on a patient in his circumstances. As Peter put it, 'I wonder if one day the burdens of living a worthwhile life might come to outweigh the wonder of it.'

On his last trip to the States in August 2007 Peter gave a revealing interview to the *Washington Post*. He admitted that his artificial heart had precipitated some religious crises, causing him to question his Catholicism. Questioning the very afterlife he wrote, 'Who knows? These are only priests. They are not very good at being challenged on the subject.' He went through bouts of clinical depression and was prescribed anti-depressants for eighteen months, but never took them. He stated that 'several times I thought it would be better if I wasn't here. Let everyone else get on with their lives. I felt I'd like to put an end to it. But choosing the method put me off. I felt cowardly about killing myself.' He talked to a psychiatrist about suicidal thoughts:

> He wasn't too worried, it's a perfectly rational response to a difficult set of circumstances. He wasn't surprised by it. He advised me to try and think about what I was doing without trying to put me off. He challenged me – are you sure you mean it? I did mean it, but not sufficiently to overcome my fear of the actual process.

My endearing cyborg was drifting in no man's land. Seven and a half years after the implant we were way into unchartered territory. Previously no one had survived with a mechanical heart for more than four years. Peter said, 'The procedure lands you in a position that no one has ever endured – what life on a battery does to you as a person. You're an invented entity trying to cope with it, trying to deal with the emotional context of it. You become cold-hearted.' He admitted that he now had a careless attitude towards money – 'You don't care if you've overspent on your credit cards or not. If you don't have any time left you might as well enjoy it. You think, what the hell, if I want something I'll have it.'

Much of the money Peter raised charitably he spent on visits to international conferences, and at these he was a revered figure, a driving force behind the new technology. Yet the last paragraph in the *Washington Post* piece was revealing:

> Things get normalised. You no longer see yourself as odd or outside the norm. Being snatched from the brink of death and transformed into a poster child for cyborg life, despite serious psychological transformations, has been quite an experience. A roller coaster. Better than being dead, I think. Three days out of five.

By now Peter had a job with the Birmingham Settlement to help homeless and deprived people. At the same time, he was working to establish a spiritual retreat in the mountains of Wales, had undertaken a ninety-one-mile charity walk, and hiked in the Swiss Alps and the American West. Our 'dead man walking' lived for almost eight years after his operation. As a result, the United States and many European countries adopted these miniaturised rotary blood pumps as an alternative to heart transplants. Many patients went back to work. Sixteen years later, with all our accumulated expertise in managing pump patients, we're on course to achieve equivalent survival times between mechanical hearts and heart transplants.

Peter died within weeks of the *Washington Post* article. I was away in Japan, working to introduce ventricular assist devices in a culture that did not accept transplantation. His death was not pump- or heart failure-related. He simply suffered a profuse nosebleed that caused his single diseased kidney to fail. He could easily have been put on dialysis – we'd previously dialysed him for a week after his first operation – but the local hospital declined to intervene. Without treatment, the levels of potassium and acid in his blood caused his own heart to fibrillate and the pump was switched off. Had I been in the UK we'd have taken over his care and treated him. I regarded this as a wholly unnecessary death.

We asked Peter's wife Diane for permission to perform an autopsy so we could investigate the long-term effects of pulseless circulation. The pump itself was pristine – there was no blood clot and minimal wear on the rotor bearings. We returned it to Rob Jarvik in New York, where it has continued to work for years on a bench testing apparatus. Peter's own left ventricle remained hugely dilated, still functionally useless. The only finding related to the pump was the thinning of the muscle layer in the wall of his aorta. As he'd had little or no pulse pressure his aorta didn't need the thickness of muscle that the rest of us have, a perfect example of how nature adapts to circumstances.

Peter left an important legacy. His experience confirmed the enormous potential of mechanical blood pumps to provide a good quality of life for the many thousands of patients with severe heart failure who are not eligible for a transplant. There are few if any ethical dilemmas, however hard you search for them. The reality is that patients targeted for this treatment have short, wretched lives.

Peter made it clear that extra life is not ordinary life. There's a price to pay and a second dying to come. But he was the first to reveal the true potential of blood pump technology, and I was pleased to play my part in something that most people believed to be impossible. He was a truly remarkable man.

11

anna's story

Body and mind, like man and wife, do not always agree to die together.

Charles C. Colton

MY WORK IS TO HELP OTHERS at the most vulnerable stage of their lives – after they've discovered that they have a serious heart problem. When they meet me it's clear to all of them that they could die, indeed some expect to. One lady was so certain of it that she made it happen after a completely straightforward operation. Never underestimate the human mind. It's a powerful thing.

One thing is for certain. For the patient and family, every single professional contact is emotionally charged. This was never more the case than for Anna. She'd had a difficult start in life. Her mother died when she was just eleven months old, but she was fortunate to have two other strong characters in her life. Her father David brought her up in a peaceful Oxfordshire village close to the church – and not just geographically – and later on her husband Des supported her through thick and thin.

Seven months after Anna was born her mother suffered an extensive stroke. It came completely out of the blue in her mid-thirties, and why it happened at such a young age was never explained. It was the last contact she had with her baby daughter. When David was told that his wife was dying he went straight home to wash the nappies.

Yet Anna recalls a happy childhood – holidays in Yorkshire and Guernsey, Sunday afternoon walks, picnics and outings. David taught her about nature and the great outdoors, where she discovered an affinity for birds and plants. She worked hard at school, but religious and social activities in the village were much more appealing than books. Above all, she adored small children and babysitting. In church she was the one who held the new babies, as well as ringing the bells – a longstanding family tradition.

Like my own mother, Anna left school and became a bank clerk. She started early in the morning and often worked late. She put her heart and soul into everything. As her father put it, 'Anna's inner strength and perseverance most likely came from my influence, and I'm proud to accept it is so.'

Anna met her husband Des when he was out walking his dog in the village. They fell in love, got married in July 1994 and bought a house. She was twenty-five, and was happy both at home in the village and at the bank.

Then, very suddenly, seven weeks or so after their wedding, she started to feel tired, at times absolutely worn out. She put it down to the long hours at work. Then there were inexplicable bouts of sudden severe breathlessness that were attributed to panic attacks. Out of the blue a sore red spot appeared on her toe. It blistered and became infected, and although antibiotics fixed the infection she wondered what could have caused the blistering, which still remained. Unbeknownst to her at the time, these were the classic symptoms of a rare and life-threatening condition, the same one that her mother had suffered from. But no one took the trouble to find out, and life took over.

At 9 am on 29 August 1994 Anna was in bed nursing a violent headache. Not a hangover – she didn't drink. Des was reading the newspaper downstairs and she recalls that *Skippy the Bush Kangaroo* was on the television. The room suddenly started spinning, and she felt that she was losing her grip on reality and going to a strange and different place in her head. She just managed to shout out to Des downstairs to call the doctor before everything went black. Anna could hear Des on the telephone and the anxiety in his voice worried her. She felt she needed an ambulance. Her brain knew what she wanted to say but her voice and mouth wouldn't cooperate. It was as if her brain had been separated from her body, which was now lifeless and unresponsive. The experience was terrifying for her.

Anna was rushed directly to the John Radcliffe Hospital in Oxford, where she appeared to be unconscious and paralysed on arrival. The paramedics wheeled her straight into the resuscitation area. 'Airway, breathing, circulation' is the rescue mnemonic, the medical ABC. Every doctor, nurse and paramedic learns this.

The doctors passed a tube into her windpipe to prevent her choking to death, then started to breathe for her with the mechanical ventilator. Her pulse was steady and strong, her blood pressure elevated – high blood pressure that goes with brain injury. So the circulation part was fine. Or was it? Did anyone listen to her heart or notice her blistered toe? Was her mother's death factored into the equation? To be fair there had been no time to look into the family history yet. It was a matter of saving Anna's life first, then afterwards establishing the cause of the catastrophe.

Diagnosis is like a jigsaw puzzle. You need to find the pieces first, then fit them together. Only then does the full picture emerge. Anna had clearly suffered a sudden catastrophic brain injury. In young people this is usually caused by a bleed into the brain from a congenitally weakened and ruptured blood vessel.

But there's a second possibility, an event known as a paradoxical embolism. An embolus is a piece of foreign material floating around in the bloodstream; broken bones can release globules of fat from the bone marrow and blood clots can detach from a deep vein thrombosis in the legs then float to the lungs. Should air enter the circulation through a cannula and drip it can block blood vessels leading to the brain or cause an air lock in the heart. A paradoxical embolism is where a blood clot breaks off from the veins in the legs or pelvis, but instead of floating to the lungs it passes through a hole in the heart to reach the brain, which can cause a sudden and sometimes fatal stroke. Anna needed a brain scan with a view to emergency brain surgery. There was one positive sign, however. Her pupils were normal size and reacted to light. She was not brain dead.

The brain scan was performed with injected dye to show the arteries within the brain. This reveals its magnificent architecture, like the branches of an oak tree – but a tree of life with a branch sawn off, as one blood vessel had come to a premature stop, although there was no bleeding. It was an embolus lodged in a critical artery supplying the brain stem, cutting off flow to that vital nerve centre.

A crucial mass of white matter was already dead or damaged, including the nerves to the arms and legs, the nerves controlling speech, together with those that govern the body's automatic reflexes. She appeared to be deeply unconscious, and probably blind.

Yet how could Anna be able to hear and think when she appeared to be completely gone? This is something straight out of a horror movie, like being buried alive in a coffin with a window – the dreaded 'locked in' syndrome, involving the complete paralysis of the voluntary muscles in all parts of the body except those that control eye movement. And worse still, only vertical movement of the eyes and blinking remain. Yet there's no damage to the thinking brain – the cerebral cortex or grey matter – and the patient remains alert and fully conscious, still able to think but mute and immobile. It's a nightmare scenario.

Anna never did lose consciousness. Her vocal cords were not paralysed but her ability to coordinate breathing with speaking had gone. So while to the outside world she seemed to be in a deep coma, from Anna's perspective her hearing and thought processes carried on as normal. Understandably, this trapped new life was terrifying to her. She could see out – although those around her were complete strangers – and she could hear a persistent, intermittent beep, which was the monitor. As her nervous system lost control she felt cold inside, even though she was covered in warm blankets. It was as if her body had been frozen and strapped down.

She recalled an olive-skinned man in a green top and trousers who was trying to put a tube into a vein in the back of her hand. He seemed to be digging about and it hurt. She couldn't move a muscle or make a sound, but she was screaming inside. He didn't speak to her, and it was as if they were in completely separate worlds. Anna wondered if she was dead but being experimented on. Where was God or Heaven now?

And where had the embolus come from? If it had originated from veins in her leg there would have to be a hole in the heart to let it through from the right side of the heart to the left. Many healthy people have a small hole between their right and left atria that's left over from the foetal circulation in the womb; it's there to divert blood from the right heart to the left before the lungs expand at birth. Anna needed an echocardiagram. Indeed all stroke patients *should* have one, not least so further similar episodes can be prevented by closure of the communication.

Anna's scan told the story, linking her own condition to her mother's early demise. There was a huge tumour filling her left atrium. Although it was fragile looking, like a delicate piece of seaweed, it was forced into the mitral valve every time the atria contracted, effectively obstructing the left side of her heart. This would explain her breathlessness and tiredness.

Her infected toe also began as an embolus, a small piece of the fragile tumour having broken off as it pounded into the valve. The next fragment went north, not south, directly through the carotid artery to the basilar artery and the brain stem – a catastrophic route that a self-destructive sat-nav couldn't have selected any more accurately.

I'd operated on many heart tumours, rare though they tend to be. Anna had a myxoma, which are common but benign. They're often fragile like hers, so bits break off. Many cause a stroke as the first symptom and are operated on immediately upon discovery for this very reason. Fortunately, most myxomas never come back after removal.

The cardiologists were called to see her. Dr Forfar wanted me to remove the tumour as a matter of urgency. I was moved by Anna's story and by the sight of her lying paralysed in bed. Her eyes were open, with a blank stare – no movement, no response. Ironically, when I held a stethoscope to her chest I could hear the murmur of her obstructed mitral valve and the 'plop' of the myxoma into the orifice. Had no one listened to her heart before? At that stage we didn't really know about her neurological prognosis. We tend not to operate on patients after a recent stroke as the anticoagulation for the heart–lung machine can cause more bleeding into the brain, but then

again there was the very real risk that soon more fragments of her tumour would embolise and prove fatal.

It was really a decision for Des and David, Anna's
husband and father. Did they want me to do it even if the
prognosis was poor? This was very difficult for them –
they were shell-shocked, and David had already lost his
wife; now his precious daughter was in the same situation. They both wanted Anna to have a chance. What did
I think, they wondered. I felt there was absolutely nothing
to lose by operating, so when they decided that we should
go ahead with the operation I took her to theatre that
same afternoon.

Anna had a small, vigorous heart that was beating
away and looking completely normal from the outside.
On the inside, however, it was a land mine primed and
ready to explode. It was important not to touch it and
disturb the delicate fronds of tumour before their escape
route could be blocked by a clamp across the aorta.

First we went onto cardiopulmonary bypass to support
the circulation and empty the heart. Next I applied that
clamp to halt blood flow to the coronary arteries, with
cardioplegia solution stopping it altogether. With the little
heart lying flaccid and cold, I opened the right atrium.
Heart surgery is simple – or it should be.

The myxoma was attached to the other side of the
partition between the right and left atria known as the
atrial septum. The safest way to approach it was to cut
away the septum and locate the base of the myxoma.
There's often a short stalk between the septum itself and
the mass floating in the blood, the aim being to remove

the whole thing so it can't grow back. This is best done in two steps – cut the stalk and lift the fragile lesion out gently without breaking bits off, then excise the whole base, which is precisely what we did. I proudly dropped the tumour into a container of formaldehyde preservative, a present for the pathologist to check that there was no malignancy. I'd operated on patients where the benign myxoma grew back and turned malignant. Rare, but it can happen.

With the tumour gone, Anna's heart separated easily from the bypass machine and we closed her up, leaving her badly wounded but safe from further damage. The surgery itself was not the greatest challenge. As a quadriplegic patient Anna's ability to get over the operation remained in question. She couldn't respond to instructions, and we'd no idea if she could breathe independently or whether she could cough. To lie flat and immobile is a recipe for chest infection and pulmonary embolism from thrombosis of veins in the legs.

We'd have to work hard to bring Anna through this journey, and as well as us it was a task for the physiotherapists and her friends and family. They were encouraged to talk to her and play her music, even though she gave no sign of being aware of anything whatsoever. When Des put earphones on her head to play her music from the local radio station there was no response at all.

Remarkably, however, Anna *was* aware of everything around her. As the anaesthetic drugs wore off she could see and hear again but still couldn't move. Worst of all, she felt pain that she was unable to communicate. To any observer in the outside world she remained in a deep coma.

One night when Anna was lying there sweating, a new nurse changed the sheets on her bed. In kindness she stroked Anna's head and said, 'I'm sorry I can't do anything more for you.' Anna panicked inside, taking this compassionate comment to mean that she was dying. On another occasion a less sympathetic nurse said, 'She looks dead!'

One day two nurses were changing the bottom sheet on her bed. As they rolled her from side to side Anna's recurrently dislocating right kneecap displaced, although no one apart from her recognised the fact. Leaving it dislocated was agonisingly painful for her but there was no way she could let anyone know. Eventually an observant junior doctor spotted the strange asymmetry of her knees and put the patella back in place. No anaesthetic. Nothing.

Des and father David visited every evening after work, hoping to see signs of improvement. I passed by her intensive care bed several times each day as it lay on the route between my office and the operating theatre. My immediate thought was that she had severe and irrecoverable brain injury. But I'm no brain doctor.

Anna's uncle visited on the evening of Monday 5 September and, like everyone, sat there trying to talk to her. The tape that kept her eyelids down to prevent the surface of her eyes from drying out had been removed. Suddenly Anna opened her eyes and her uncle jumped up in surprise, shouting, 'She's awake, she's awake, Anna's awake!' Not only that. She could follow the movement of a finger up and down with her eyes. This was the first indication of consciousness since her stroke a week before.

Des and David had just left the hospital, having been there most of the day. When they heard the news they rushed back but by then Anna had fallen asleep. With the realisation that Anna was not brain dead it was reasonable to let her try to breathe for herself. Over the next twenty-four hours we managed to remove the breathing tube from Anna's throat, a great relief for her, and making physiotherapy and bed changes easier.

A few days later Anna was awake for most of the day, breathing well, with a stable pulse and blood pressure. There was the usual pressure on intensive care beds and, against the family's wishes and my own grave reservations, she was transferred to a single room on the ward. With less frequent chest physiotherapy she soon developed pneumonia, which needed treatment with a combination of antibiotics. Still prostrate and unable to cough, this developed into a life-threatening situation, a high, swinging temperature, profuse sweating to the point of dehydration and uncontrollable bouts of shivering making her life intolerable.

The pneumonia was getting worse not better. Then by chance Des saw the letters 'DNR' scrawled across the cover of the brown folder containing her medical notes – Do Not Resuscitate, written on the grounds that her projected quality of life would be unacceptably poor and without any permission from the family. It gave them all a clear message that the medical staff had given up.

What it meant specifically was that Anna wouldn't be put back on the ventilator if her chest infection proved overwhelming. David said about this, 'I think it was put on her records when she was moved out of the intensive care unit. I'm not certain of the ethics but felt they should have discussed it with us.' Of course they bloody well should have. Vets don't let pets die without discussing the issues with their owners, and it would have been reasonable – to put it mildly – to mention it to her family. Scary.

Now that Anna was in a single room on the ward she was solely my responsibility, not the intensive care doctors'. I called a case conference with my own surgical assistants, the ward nurses and the physiotherapists, then I brought in Des and David for a frank and open discussion. We'd come so far with Anna – she was awake and, although the prospects of neurological improvement were limited, the family wanted her to have the best chance.

What did 'Do Not Resuscitate' actually mean? With the myxoma gone she had a normal young heart that was never going to stop, and no one was going to have to pound on her chest or zap her with the defibrillator. What she needed was physiotherapy and antibiotics for a while, together with loving care to make her feel human again. In no way was she just an inconvenient object in a bed who needed a bit more effort than usual. The pep talk served its purpose, the team pulling together and curing her of the pneumonia.

Gradually Anna remained fully alert for longer periods and was soon sat out in a chair. Her breathing improved and she learned to communicate by blinking, making yes or no responses to questions. Well-meaning nurses worked out a system for others to communicate with her through winks and blinks, but unhelpfully they taped the instruction sheet onto a locker too far away for Anna to see. And no one thought to put her glasses on. In time she regained some ability to control her head movements, then she learned to use a specially devised 'speech board' to interact with visitors. The process was slow but began to give her a means of expressing her well-preserved intellect. In time she began to tell us her own story, the things that she remembered from the other side of the fence.

I remember waking up in what I assume was the middle of the night. It was very dark. There was that intermittent beep all the time and what looked like lots of televisions lit up. I now know they were the heart monitors in intensive care. It felt like my neck was resting in a bowl. Someone was pouring lovely warm water over my hair and massaging my whole scalp. Whoever it was they were washing my hair! It felt absolutely wonderful.

When they had finished, the bowl was taken away and I tried to hold my head up. I wanted to see where I was. My neck seemed to have lost all its strength and the back of my head felt as if it had been filled with concrete. I couldn't speak and don't remember being able to cry at all. I was frightened. There were curtain rails in a square above me and a painted ceiling. Unable to move or lift my head, I just lay flat on my back looking directly upwards. No sign of life in my vision but plenty of voices. One voice I recognised. A woman. My line manager at the bank. I was worried she had come to check up on me. To see why I was not at work. One person mentioned a funeral the next week. I thought it was mine. My uncle realised and reassured me. My brain worked fine. Where was my body?

Often lots of people in white coats gathered around the bed. Always talking about me, not to me. Things I'd never heard of before. Then they just left. I wanted to ask them stuff. Where was I? Why was I here? How dare they talk as if I wasn't there? I was indignant but I couldn't get it across. Much of my confused state and dreadful thoughts could have been prevented if people talked to me. No one explained what had happened to me.

One day a medical registrar called Imad came to see her from the Rivermead Rehabilitation Centre. He was kind and actually spoke to Anna. He asked if she would like to have the feeding tube through her nose replaced by one inserted directly into her stomach.

'I hated the tube in my nose,' Anna recalled. 'I opened my eyes wide and smiled to indicate "Yes". That was the first time I remember anyone trying to involve me in my own care.'

Imad was there to assess Anna for a rehabilitation programme when she was fit enough to leave hospital. This was still three months away, as she needed to be much stronger and able to swallow before she could leave. Progress was slow but steady. She had a few more chest infections and further courses of antibiotics. At least they removed the Do Not Resuscitate order from the front of her notes. Anna was very much alive – and wished to stay that way. By the end of January she was sufficiently strong to move her head and blink, and was in a fit state to move on. Although she remained quadriplegic, being able to breathe without a ventilator was a great blessing.

In all, it took almost three years before Anna could move back – in the Easter of 1997 – to her modified home with Des and start rebuilding her life. She remained physically dependent but mentally alert. On weekdays Des went off to work early, then two helpers would arrive. They got Anna out of bed, after which one stayed with her for the morning. At lunchtime a different carer came until around 7 pm. Then two others would arrive to help put her to bed. A rigid routine. She got out and about to the supermarket and local park in a sophisticated electric wheelchair directed by movements of her head. She liked to be treated as a normal person and have people talk to her.

From a scripted box of tricks mounted on the wheelchair Anna could open and close the front door, draw the curtains and operate the television. The box was programmed to operate with an infra-red controller. A nod pushed against a lever on the left side of her head and this triggered a cursor that stepped down a list of commands. When it reached the chosen instruction she nodded on the lever again to select it.

Anna also had a computer room overlooking her garden. In it a receiver monitored her head movements via a white reflective dot fixed to the bridge of her glasses. This enabled her to direct a mouse around the computer screen, and with specially created software she could write emails and keep in contact with her friends. Like predictive texting on mobile phones, her computer constantly second guessed what she intended to write.

Apart from loss of mobility, Anna professed that little had changed for her since the stroke. As a religious woman she accepted her situation and made the best of it. The local radio station ran a campaign to buy her an adapted van that could carry her wheelchair. Her father christened the blue Vauxhall Combo the 'Annamobile' after the similar Popemobile. And her main worry in life? That she might grow another myxoma that I couldn't remove from her heart. She was content in her own body and didn't want life to be cut short by another stroke.

Dr Forfar kept her under surveillance every six months with an echocardiogram. The first myxoma had been radically excised and was unlikely to return. But I was aware of genetically based familial myxomas and was convinced that Anna's mother had died from one. Patients with the familial myxoma gene can develop further tumours at different sites and I just hoped that this wouldn't happen.

But in August 1998 I received a call from Dr Forfar, who had Anna and Des with him in the office. There was devastating news from her latest scan – the myxoma had recurred. He told me that Anna was very frightened and asked when I could take it out.

I assured him that if he could bring her into a cardiology bed that afternoon I'd do it the next day. It was a reoperation, so we'd need to have blood available. Reoperations are always more complex, and in this case the sac around the heart would be obliterated by inflammatory adhesions from the first procedure. As I'd learned years ago at the Brompton, the heart can be stuck to the back of the breastbone. But I'd done hundreds of reoperations since that first débâcle and it wouldn't be a problem.

Anna was sitting in her wheelchair looking petrified when I saw her on the ward. Des was crestfallen and her father David was on his way. We'd meet in the operating theatre the following morning, and I said that everything would be fine but I needed to go and change my operating list. In reality, the emotional black hole was about to pull me down and I needed to escape.

Des came to the anaesthetic room with her. He stayed to keep her calm until she was unconscious. The first time I met Anna she was already paralysed but still had muscular arms and legs. Now, up on the operating table, I saw that these had markedly wasted away by three years of immobility. I listened with a stethoscope before painting the chest. I was convinced I could hear this one and, as I suspected, the origin of the myxoma was at a different site, towards what we call the left atrial appendage. There was no stalk – it simply had a broad base, which I chopped out, and then I sewed the atrial wall back together.

I looked carefully around the rest of the heart to make sure there were no other tumours lurking in the recesses. Nothing. We came off the bypass machine easily, closed up again and took Anna back to intensive care. We knew that she'd wake up this time, so she had her communication mechanisms ready. The physiotherapists were on standby. It was all much easier after the dress rehearsal. Once again her family and friends rallied round, and I hoped for her sake that it would be the last time I saw her.

It wasn't. The third time Anna was thirty-two, and seven years had passed since the first operation. In April 2001 her follow-up scan showed another huge myxoma in the left atrium, again in a different position – directly above the mitral valve. This tumour was more solid, and plopped in and out of the valve orifice. A dangerous situation. Large myxomas can block the valve altogether and cause sudden death. Again Anna and the family were in a state of distress, having watched the echocardiogram unfurl on the screen.

I took her straight back into hospital and to the operating theatre the following day. The third time through the sternum is always tricky. Again I entered the heart via the right atrium and opened the remains of the septum. The tumour lay directly in front of me, originating next to the mitral valve and partly from the atrial septum. I set about lifting it from the left atrium with an ordinary kitchen spoon, a useful implement for tissue that has the consistency of jelly. I'd never seen or heard of a patient needing surgery for more than three heart tumours, and soon we'd run out of places on her little heart to insert the bypass cannulas.

Anna bounced, or rather crept, back again. Her spirit and the support of Des and David were extraordinary. She suffered the inevitable chest infection but the physiotherapists brought her through. We were assiduous about pain control and used the same communication mechanisms as before. This was the benefit of a consistent ward nursing team in those days.

She spent three more weeks in hospital before going home. We learned that she'd been struggling with depression, but it would be inconceivable for this not to be the case – massive stroke and multiple heart operations, then the realisation that this must have been the cause of her mother's premature death, and worst of all the persistent anxiety about the tumour recurring. It had come back twice at different sites, so would it happen again, and was a fourth operation indeed technically feasible? Could it be done safely? We all hoped that it wouldn't come to that.

Now Des couldn't bring himself to attend the follow-up appointments. The strain of sitting there watching the echo images on the screen was just too much. Instead he went to the church to pray. Anna was painfully thin, so her echo pictures came out as clear as day, and on each follow-up she lay there, desperate to see empty chambers, atria that were getting smaller with each operation.

In August 2002, just sixteen months after the last procedure, came another nasty surprise. Dr Forfar called to show me the monster – the largest tumour so far. I didn't believe that a new myxoma could have grown to that size within months. I said nothing, but wondered whether this one might be malignant. I'd operated on a young woman with that scenario before. The first myxoma was benign, the second time it was a highly malignant myxosarcoma. We didn't want that for Anna. I brought her back into hospital for an urgent fourth operation.

To obtain written consent for an operation we're obliged to explain the risks. No one could claim that the risks of death during a fourth heart operation could be any less than 20 per cent. Equally there were significant risks of a further stroke as there was a real chance that bits of myxoma would break off and visit the brain. But if we didn't operate the tumour would continue to grow rapidly and obstruct the heart. The bigger it became, the greater the danger of embolism. We were stuck between the Devil and the deep blue sea, and I figured that we'd be better off taking on the Devil. Anna and the family would have divine help in that struggle. Plus she couldn't swim.

On the day of Anna's surgery the church held a vigil for her. As always Des and David brought her round to theatre. I stayed in the theatre coffee room. For them it was an emotionally draining experience, just like when parents bring a young child to the anaesthetic room and have to leave it with strangers. Des had doubts that she would get through this one.

This was the biggest and most aggressive myxoma so far, practically filling the whole left atrium. I did a radical excision, then took a long, hard look at this battle-scarred cavity. Was there anything I could do to stop further growths? I decided to take the electrocautery and kill the

cells of the whole inner lining, the layer that was genetically programmed to terminate Anna's life. I fried as much
as I could see, the smoke rising like when stubble's burnt
in a corn field. I'd decided upon a scorched-earth policy
as I simply didn't want Anna to succumb to this curse.

As we obliterated the cell lining I had an extraordinary
and unexpected stroke of luck. I'd pushed open the mitral
valve to look around the left ventricle and spotted a baby
myxoma on one of the muscles of the mitral valve, too
small to be seen even on the best echo but destined to
grow large had we missed it. Out came the bastard and
into the pot it went with the rest. Everything had to be
analysed by the pathologist.

The heart continued to look good in normal rhythm
and frying the inside of the left atrium had provoked no
adverse effects. I watched the rest of the operation over
the drapes – my team was first class, so I wasn't needed to
finish off.

When the sternum was wired together I went to call
Des, wanting to put the family out of their misery as soon
as was reasonable. There had been very little bleeding, so
the whole procedure had been quicker than anticipated
and I suspected that he might still be in the church. When
I got through I told him the battle was over. Once again
Anna was safe within the bounds of her ability to bounce
back from the trauma of it all.

But I was concerned that she might give up if she dwelt
upon the recurrence issues. Anna needed an overdose of
positive thinking, a massive morale boost to get her
through the next few weeks and carry her past the pain,

the fear and the uncertainty. So I asked Des to bring God back to the hospital with him.

Anna recovered slowly, this time without a serious chest infection. Everyone rallied round once more to see her through – medical staff, nurses, physios, the chaplains, and especially her own friends and family – all with a gargantuan dose of positive thinking. By now she was a well-known character in the hospital and the community, and everyone was willing her better.

Once again she went home, only to face the inevitable outpatient's appointments and the much-dreaded echo-cardiograms. Months passed without incident. Then years. Two of them, at least.

Then came the grey, wet November afternoon the day before Guy Fawkes Night in 2004, and Anna's routine appointment with her cardiologist. She was with her dad, who helped her onto the couch for the echo. Gel was applied to her bony little chest to improve contact with the probe and both of their adrenaline levels rose in anticipation. But within seconds came that familiar sinking feeling as they saw yet another lump floating around in the left atrium like a goldfish in a jam jar. So much for my scorched-earth policy.

This was too much for Anna, too much for David – and Des – as well. It was easy to understand their standpoint: how much could one person endure, why was God letting this happen and, more to the point, where to go from here? This last question needed careful consideration. How much heart could be removed from this young woman? The situation was too emotionally charged to make a quick decision, so Anna and her dad went home in abject desolation. Dr Forfar had to think about it as well and would discuss it with me, but he let the family settle down for the Christmas period. Of course they couldn't rest. Peace of mind was impossible, as Anna knew this was a death sentence.

She returned with David in early February, and Des came too this time. There was no uncertainty for him now, only the discussion about what could be done – if anything. A repeat echocardiogram made them unbelievably miserable. All of Anna's four myxomas to date had grown rapidly, although they were benign. This new one was 2 cm in diameter and already prolapsing dangerously through the mitral valve, so a further stroke was likely.

Dr Forfar called me with the dismal news. What did I think? Would Anna be considered for a heart transplant? Sadly not. A transplant leaves a large cuff of left and right atrium to which the donor atria are sewn, so it wouldn't protect her. A heart and lung transplant could remove the whole heart, but no one would consider that because both lungs were stuck to the chest wall after the previous surgery. I said I was willing to operate again, but we all needed to agree that it would be the last time. Between the two of us we felt that we couldn't just leave Anna to the inevitable.

When asked, the family agreed that she'd rather die during the surgery than be abandoned. In the event of success there would be no further echocardiograms. A head-in-the-sand strategy, admittedly, but there was no point in making everyone miserable again.

Anna was admitted on Valentine's Day, eleven years to the day after she and Des were engaged to be married. This fifth operation was predictably difficult and dangerous. With patience and great care, we re-entered the chest and dissected out just enough heart to get back into the right atrium. Having achieved that safely I went out for a break. This is a good strategy in complex reoperations and a necessary one for surgeons with an ageing bladder. Now for round two.

I opened the right atrium to approach the left, intending to go directly through the patch from operation three. At the mouth of the inferior vena cava draining from the abdomen was a completely unsuspected right atrial myxoma, as large as the one we were chasing on the left. We removed it, although in truth it almost fell out. Then we lifted out the left atrial myxoma. Job done again, with a great sense of satisfaction. We closed up the heart, removed the air and warmed the blood. Unperturbed by this fifth insult the much-abused little organ bounded off the bypass machine. Two myxomas for the price of one. Again. We closed the chest over it, never to be exposed again. Relief for me, resignation for the family.

At first the post-operative course was straightforward. Anna spent two days on the ventilator, then the tube was removed and she had frequent physiotherapy. Everyone was elated that she'd survived. Then she was given some soup without adequate supervision. With the brain stem stroke her swallowing had always been an issue, and she inhaled the hot liquid, then choked. There followed a long period on the ventilator with a chest infection, which needed several courses of antibiotics and eventually a tracheostomy. But she came through in the end and was no worse than before. Anna and Des returned home to get to grips with the uncertainty, to try to banish the depression and have the best life possible.

Time passed and we didn't bring her back to the hospital. Rivermead were very supportive and kept an interest in her. Above all she was well supported by the church and the community. From time to time I would ask Dr Forfar whether he'd heard anything, although after a while we both lost track of her and heard nothing until I discovered that a neighbour knew her well from church. Then I had serial updates. She was happy. Des was happy. He'd stuck by Anna through thick and thin. Occasionally I would receive a card.

In 2015, more than ten years after her fifth and final operation, the Annamobile pulled up outside my home. She was in her wheelchair at the back, beaming and blooming. Des came to the door with a cake. With the carers' help Anna had made it for me to celebrate their twenty-first wedding anniversary.

So what had happened to the myxomas? The genetic storm had abated and the battle was won. With divine help I expect. It brought to mind a line from 'The Flower' by the seventeenth-century poet George Herbert: 'Who would have thought my shrivel'd heart could have recovered greenness?'

I hope they both live happily ever after.

12

mr clarke

*Before you tell the truth to a patient, be sure
you know the truth and that the patient
wants to hear it.*

Richard Clarke Cabot

18 MARCH 2008. I was ambling back to my office after
the first case of the day – a baby with a hole in the heart;
nice result, and happy parents – when I saw a woman
weeping at the far end of the corridor. She was smartly
dressed, with two young children holding on to her coat.
Although it was none of my business, after forty years in
surgery I was still not immune to other people's grief. So
the desperate little tableau upset me.

Everyone else strode past them in a purposeful way,
going about their hospital duties – nothing to do with
humanity or common decency, more about deadlines,
figures or waiting lists. I was about to divert towards my
office and a pile of paperwork, but I couldn't do it. Even

though I looked and felt a mess in my sweaty theatre gear, I approached her.

The poor lady was so consumed in grief that she didn't notice me at all, or if she did she must have thought I was a porter waiting for the lift. Quietly I asked whether there was anything I could do. After a minute passed in which she tried to compose herself, the lady explained that she'd left her husband in the cardiac catheterisation laboratory. He was dying and they'd been told that nothing more could be done. Now she needed someone to look after her children, then she could go back and sit with him so that he didn't die alone.

I pressed her for more information. Her husband, Mr Clarke, was forty-eight. Earlier that morning – and without any warning – he'd suffered a massive heart attack. First he was taken by ambulance to the nearest district general hospital, where he suffered a cardiac arrest and was resuscitated and placed on a ventilator. Having established the diagnosis of myocardial infarction, the cardiologist inserted an intra-aortic balloon pump and relayed him on to Oxford – more than an hour away – for urgent angioplasty.

The objective of angioplasty is to open the blocked coronary artery and stop the oxygen-depleted heart muscle from dying – the infarction bit of myocardial infarction. The cardiologist feeds a balloon catheter through the aorta and into the blocked coronary artery, and it's inflated to open up the tiny vessel, a small metal stent being inserted to keep it open. In most cases this reinstates blood flow to the compromised heart muscle by a process known as reperfusion. Now the critical bit: reperfusion within forty minutes of the onset of chest pain salvages 60 to 70 per cent of the muscle at risk. Beyond three hours, only 10 per cent will survive.

Mr Clarke had been bounced from pillar to post, his treatment taking much more time than was reasonable. The treatment guidelines advise the use of 'clot busting' drugs in cases of delay. These can dissolve the blood clot that's blocking the narrowed artery and should restore blood flow – not as good as angioplasty, but better than nothing.

Oxford has a fantastic emergency angioplasty service. It's round the clock, all day, all night. Once in the cath lab, Mr Clarke got the best treatment. His blocked artery was opened, but the left ventricle – badly damaged during the delay – now wasn't moving and there was very poor blood flow. A normal heart pumps five litres of blood per minute, while his heart was managing less than two. With a low blood pressure of around 70 mm Hg, half of what's normal, lactic acid was accumulating in his blood. He'd reached the stage we call cardiogenic shock and was sinking fast. Without a miracle he was buggered and his kids would lose their dad.

I didn't want that to happen and told Mrs Clarke that I'd see if there was anything I could do to help. Maybe we could try one more thing. Because of our past achievements I'd been sent a new ventricular assist device to test from America. It was time to give it a try!

We agreed that Mrs Clarke should take the children to the cafeteria to try to divert their minds from the misery and I'd come back to them. I needed to get Mr Clarke into that operating theatre as soon as possible and would have to reschedule the operating plan for the day. We'd start by supporting him on the heart–lung machine to improve his life-threatening metabolic state, then we'd take over from his dying heart.

I started to make my way down to the cath lab past my Portakabin office. My new secretary Sue was killing ants on the windowsill, still waiting for me to get to grips with my paperwork. Mercifully there was a new excuse for me to avoid it. I asked her to call the anaesthetic room of Theatre 5 and warn them about the change in plan.

'What plan?'

Sue was quite entitled to ask this, as she'd no idea about Mr Clarke, but there was no time to explain. And could she please warn the perfusionists that I was going to use the new CentriMag pump.

I wanted to see the coronary angiogram so I knew what we were dealing with and whether the heart stood a chance of recovery. This only took two minutes. The left anterior descending coronary artery had been completely blocked but was now once more wide open with a metal stent through it, preventing it from closing off again. The coronary flow wasn't as brisk as it should have been, and the echo showed a substantial part of the left ventricle was indeed motionless and not contracting at all, even though the artery was open.

The $64,000 question was whether the muscle was already dead – myocardial infarction – or whether it was suffering from what we call 'myocardial stunning', which, while bad, was not nearly so serious. 'Stunned' muscle remains alive but takes days or weeks to recover. We'd find out if I succeeded in keeping him alive.

There was no chance to explain all this to Mr Clarke, as he was very quickly going downhill. He had the ventilator tube down his throat lying flat on the trolley, and when I tried to introduce myself it was clear his mind was failing, bordering on unconsciousness. His kidneys had stopped producing urine, his lungs were filling with fluid and he was icy cold, deathly pale yet sweating. There was froth in the corner of his mouth, bubbling through blue lips, and his eyes were rolling. This is how heart attack patients die, and was how I lost my grandfather. There was no time to send for porters so I asked the nurses to head for the lifts. Just get him up there before he arrested. I'd deal with the consent form later – whether he lived or died, he certainly wouldn't be suing me.

They say everything in life is about timing. In Mr Clarke's case, timing was the stuff of fantasy – you couldn't make it up. My chance encounter with the distressed lady on the corridor. An empty operating theatre. And the new CentriMag pump. It was reminiscent of Julie's good fortune with the AB-180. They were the lucky ones.

The pump was called a CentriMag for a good reason. The blood propulsion mechanism – known as the impeller – spins within a magnetic field like a centrifuge at up to 5,000 revolutions per minute. Centri – centrifugal, Mag – magnetically levitated. It can pump up to ten litres of blood per minute, far more than needed. Limited pumping capability had been the drawback with artificial hearts from the outset, but now the technology was improving rapidly.

Mr Clarke, by now a metabolic wreck, was too sick to linger in the anaesthetic room, so he was wheeled directly through to the operating table. To give him a general anaesthetic at that point risked immediate cardiac arrest – instead, the monitoring lines and transfusion cannulas were inserted under local anaesthetic. To keep him alive I had to get him onto the heart–lung machine rapidly, then his blood needed filtering before we switched to the CentriMag system.

The sternotomy incision was bloodless. Corpses don't bleed. The injured heart quivered, giving up the ghost, but as always cardiopulmonary bypass changed everything. The struggling heart emptied and I had a good view of the stiff muscle that had been starved of blood and oxygen. It was clear that it wasn't dead, and I could even see and feel the coronary stent sitting within the artery like a rat in a snake's gullet, blood coursing through it to the swollen muscle. The ventricle was down but not out.

Mr Clarke was experiencing a bog-standard death from heart attack, the sort that happens to hundreds of patients every day across the NHS. I had a grim determination to show that he could still be saved with the right technology. For the sake of that family.

With the CentriMag system, plastic tubing diverts blood from the left atrium out of the body to an external rotating pump head, then more tubing brings blood back to the chest and into the aorta, where it emerges from the heart. A control console the size of an old-fashioned type-writer regulates the pump speed. This simple arrangement bypassed Mr Clarke's struggling left ventricle and allowed it to rest, at the same time providing generous blood flow to his brain and body.

By releasing clamps on the tubing we allowed it to fill with blood, which pushed out air. As always the whole system must be airless. It was our obsession – that little saying, 'Air in the head, dead on the bed', couldn't be repeated often enough. Now it was time to switch on the CentriMag. We balanced the reduction of flow in the cardiopulmonary bypass circuit with an increase in the machine's system, and then it took over altogether, just like clockwork, in a smooth and effortless transition. Magic.

I looked at the clock. It had been almost three hours since I'd dispatched the grief-stricken family to the cafeteria. Shit. They'd now be sitting there wondering whether he was alive, most probably expecting him to be dead. I felt concerned for them but there was nothing I could do about that now. Good news would make up for it.

For once I got on and closed the chest myself, taking care to protect the life-preserving tubes. By the end there were two pacemaker wires and four plastic tubes emerging from beneath his ribs, two of which were just drainage tubes to let out blood.

I went to find Mrs Clarke. By now other family members had arrived to take the children away from the hospital, and I wanted to bring her to the bedside myself. When we both returned it must have felt a bit like being in a spaceship for her – wall-to-wall technology, the ventilator to breathe for him and his circulation supported by the CentriMag. What little space remained around the bed was taken up by monitoring equipment and drainage bottles. Amid all this was her husband's broken body, something to look at rather than communicate with.

Her first reaction was alarm – the sight, after all, was an emotional stab wound – and I thought her legs would give way. We moved quickly to sit her down by her husband. Her immediate instinct was to hold his hand. He didn't respond, but at least he was warm and even pink now, not cold and clammy like when she last saw him, with that greyish blue colour of those dying from cardiogenic shock. The nurses were kind. They scraped Mrs Clarke from the ceiling, then started to explain all the paraphernalia to her. They were confident enough to manage the equipment and the instructions to them had been simple: don't change anything. We were winning.

After a week Mr Clarke's uninjured muscle looked much better, so I decided to take the optimistic route and remove the CentriMag. We returned to the operating theatre, where we slowly reduced the pump flow and watched his heart's performance on echo. The left ventricle was ejecting well, he had a normal heart rate and adequate blood pressure. There seemed to be little residual damage from last week's catastrophe. Bloody brilliant, I thought.

We took out the pump, washed his chest out, put in clean drains and closed him up for the last time. He remained perfectly stable. After another twenty-four hours he woke up and had the breathing tube taken out. He'd returned from his week away as if resurrected from the dead. When I finally spoke to him he recalled nothing of the events, and hadn't had any 'out of body' experiences or flashbacks. Nor did he have any idea who I was or any recollection of which hospital he was in.

I wanted to be there when his kids came back – not actually with them, but away in the corner of the room somewhere, just watching as they came in to see their dad. It was certainly worth the wait. Amazingly, just a week after all this, Mr Clarke went home. Equally remarkable was the fact that at his follow-up three months later his heart looked normal. All the 'stunned' and struggling heart muscle had recovered. It was a textbook 'just in time' job.

For me the Clarke case was a watershed moment. So many patients continued to die after a heart attack, even when emergency angioplasty had succeeded in opening the blocked vessel. We'd shown that at least some of these victims could be saved with simple, inexpensive technology. It had by now become a repetitive theme.

Splint a broken bone and it will heal. Rest an injured heart and, although it may recover, it won't always. But for me the patients deserved that chance. What's more, the nurses on the intensive care unit found the CentriMag system very easy to manage. Turn it up or turn it down. We had control over the patient's whole circulation by simply twiddling a knob. It was much more straightforward than driving a car.

Now the sting in the tail. Six months after Mr Clarke suffered his heart attack the same thing happened to his younger brother, who was just forty-six. I was away at a conference. The second Mr Clarke was taken to his local hospital only to be passed on to Oxford. By that time he was already in cardiogenic shock. His family received the same message as his brother's family: that there was nothing more we can do. They searched out my office, desperate for help, but I was away, so there was none. No surgeon, no pump. His wife lost her husband, his children their father.

The first Mr Clarke had to take over their care. When I heard about what had happened I was desperately sad yet at the same time relieved that I didn't have to face that family. With age, my objectivity was fading and empathy was taking over. I was suffering for my profession.

13

adrenaline rush

We are but tenants. Shortly the Great Landlord will give us notice that our lease has expired.

Epitaph on the tomb of Joseph Jefferson,
Sandwich, Cape Cod

ALLIED FIGHTER PILOTS during the Battle of Britain thrived on adrenaline, the hormone secreted by the adrenal gland in response to stress. One minute they were relaxing in deckchairs in the sunshine, the next scrambling for their planes and soaring up into the sky, anticipating the conflict ahead, then risking sudden death.

Medical students are taught that adrenaline is the 'fight or flight' hormone, flight as in escape – not flying a Spitfire. But there are times when I have to scramble like those fighter pilots, times when every minute counts, even seconds. The call comes through that a patient with a penetrating chest injury is on their way to the accident department by helicopter or by ambulance. The entry wound is close to the heart, their blood pressure low and they need a cardiac surgeon as soon as possible. Scramble!

Sometimes simple, frustrating issues spell the difference between life and death – a set of traffic lights, a police car in front, no space in the hospital car park. I cannot speed like an ambulance and there's no blue flashing light on my car. So I drive fast and get into trouble. As a senior registrar travelling between London hospitals I was pulled over so many times that the police came up with an offer.

When you need to move fast, call 999. Explain to the operator and we will take you where you need to go. They did this on several occasions, but it wouldn't happen these days. Now they flag me down and I throw a fit. I tell them to check out the incident with the ambulance service, then escort me to the hospital. This conflict pumps up the adrenaline even more, so when I get there I'm ready to explode into action, to wield that knife.

My mobile rings at 11 pm, number 'unknown'. 'Unknown' is always the hospital. The operator says, 'I will just connect you with the Accident Department,' and I'm all ears, pissed off at being disturbed late at night but listening intently. The doctor says an ambulance is on its way from Stoke Mandeville Hospital. The patient has a high-velocity gunshot wound to the left chest and is in shock. The doctors at Stoke Mandeville had put up a drip and said, 'Take him directly to Oxford.'

I asked what turned out to be an Air Force medic how he knew it was high-velocity. Because it was a hunting rifle. Was there an exit wound? No. This had important implications for the damage inside. I knew about gunshot wounds. I'd done a stint at the Washington Hospital Trauma Center, then another at the Baragwanath Hospital in Soweto, Johannesburg, and I'd written the chapter on 'Ballistic Injuries of the Chest' for the British military's textbook of emergency medicine. I loved operating on penetrating chest wounds as they're unpredictable, each one is different and they're always a challenge.

'OK, I'm coming. Could you call my registrar? Ask him to call the theatre team in.'

I had a high-powered Jaguar in those days, before I wrote it off, and the roads were dark and empty. I could let loose on the accelerator, all the while keeping a cautious eye out for deer or foxes on the road. My mind sifted through the sparse information I'd been given. How on earth did this guy get shot late in the evening with a high-velocity rifle?

High-velocity bullets follow a predictable course when they hit the chest but they spin rapidly, transferring energy-gouging holes into the lung and generating secondary missiles – fragments of metal, shards of rib, bits of cartilage. They're usually fatal. Had he been shot at shorter range the bullet would have gone straight out the back of his chest with a large exit wound.

This unfortunate gentleman lived on the edge of a woodland shooting estate. Having just switched off the television before going to bed he heard what sounded like gunshots. Was it poachers? Despite it being a cold night with a full moon and coming up to Halloween, with patches of eerie mist lingering in the hollows, he walked down the lane to the edge of the woods and out into the fields to find out.

Suddenly a crashing blow to the chest knocked him off his feet, even before the sound wave reached him – the crack of rifle fire. There was an agonising sharp pain above his left nipple that took his breath away and he immediately felt faint, but he had the presence of mind to pull out his mobile phone and dial 999. He told the operator he thought he'd been shot and gave his location, then collapsed in mental and physical shock. Staring up at the dim stars on this moonlit night, he fully expected to die.

The assailant was in big trouble. He'd been poaching deer on his own patch and had mistaken the glint of moonlight in the victim's spectacles as a pair of bright eyes. Dropping the rifle sights down to a broader target and aiming at what he expected to be the deer's chest, he pulled the trigger. It certainly was a chest – but not of an animal, and he missed the heart by an inch. This was exceptionally lucky for both of them, as no one survives a high-velocity rifle bullet through the heart.

Years earlier, at the Middlesex Hospital, I'd saved the life of a young man shot by the police in east London. The difference then was that it was a pistol bullet – this had passed straight through his heart but a blood clot in the pericardium plugged the holes, which is what happens when the pressure in the heart falls after blood loss. But with high-velocity bullets it's a totally different story. They tear the heart to pieces, so I knew our patient didn't have a cardiac injury and was confident that I could fix the rest.

I arrived before the patient. The accident department was otherwise quiet, so a horde of medical and nursing staff were on hand, waiting to pounce. But I needed just one – an anaesthetist to insert the tube in his windpipe and secure his breathing. What I didn't want was aggressive fluid infusion to replace the blood loss. Clear fluid simply raises the blood pressure, promotes bleeding and impairs the blood's ability to clot, risking catastrophic haemorrhage.

Back then the Advanced Trauma Life Support guidelines were poor if not dangerous in this respect. Research from Washington, DC, even showed that patients with penetrating chest wounds had better survival rates when brought to hospital by private car rather than by paramedics who spent time putting up drips and pushing in cold fluid.

The ambulance sounded its siren on approach, and by now the patient's blood pressure was less than 60 mm Hg with a heart rate of 130. He was cold, pale, sweating profusely and losing consciousness, and the paramedics knew that time was running out. They reversed towards the entrance and threw the ambulance's rear doors open. Down came the ramp and the patient was rushed into the resuscitation area. I asked him his name but he didn't respond.

He was still wearing a sweaty, blood-stained shirt with a ragged bullet hole in the front. Beneath was the small skin-entrance wound, surrounded by a ring of black blood under his pale white skin, and now plugged by swollen

muscle and clotted blood. What's more, I could feel air in
the tissues under the skin, a sure sign that his major
airways had been damaged. I needed to predict the injuries inside from the site of the entry wound, and it was not
reassuring. The wound was close to the root of his lung
– the major clockwork – and over the blood vessels.
Luckily it lay a little distance away from the heart.

There were far too many cooks about to spoil the broth.
I wanted him put to sleep and ventilated quickly so I could
cut open his chest and get to the bleeding. He needed a
couple of wide-bore venous cannulas in the veins, but
there was no time for X-rays or scans. He needed treatment not investigation. As the anaesthetist put the tube
down his trachea I asked the nurses to give me a gown and
gloves, then get the chest-opening instruments ready.

Panic spread with the realisation that I was about to
open him up right there on the trolley. The anaesthetic
drugs had stolen away what remained of his blood pressure
and he was about to arrest. I had to find the bleeding, stop
it, then get some donor blood into him. Clear fluid doesn't
carry oxygen – only red cells do that, and he was short of
them. I reckoned he had three litres or more of blood
spilled into his chest cavity and a completely collapsed left
lung. My registrar scrubbed up to join me. I had the nurses
roll him onto his side, left side up, then cut away the wet,
bloody shirt with scissors. We rapidly painted his skin with
iodine antiseptic and wiped away the sticky mess.

Curiously, I spotted the bullet lying under the skin just
beneath his left shoulder blade. It must have been deflected
by the scapula bone at the back of his chest then travelled

downwards to rest in the centre of a bruise. I remember thinking that we should fish the bullet out and keep it for ballistic evidence to link it with the rifle that was fired.

With a scalpel I sliced open his chest between the ribs, from the edge of the sternum all the way round to the shoulder blade, where the bullet popped out. I kept cutting with the blade, down through the thick, pale muscle layers. In a live patient these would normally hose blood – but he had no blood pressure and there was in any case little left to bleed. As I breached his chest cavity, great hunks of clotted blood like liver slithered out and plopped onto the floor, followed by fresh liquid blood. I grabbed the large rib retractor and cranked open the chest cavity, trying to expose his injuries and spot the bleeding point.

By now one of my own theatre nurses had arrived with a powerful sucker, and I could see blood welling up from the depths. As I'd anticipated, the pulmonary artery was lacerated and air was blowing out of the main bronchial tube, so I needed to apply a large clamp across the root of the lung to control both of them. The operating theatre nurse scrabbled around trying to find me one, and once she'd done so and it was safely in place I told the anaesthetist to transfuse him rapidly.

The patient's heart was slowing down, grinding to a halt in fact. I could see it right in front of my nose through the thin pericardial sac, so I stuck my fist around it and pumped hard for a few cycles to give it some help. It felt empty. I asked for a syringe of adrenaline and stuck the needle directly into the apex of the left ventricle. A couple of millilitres would cheer it up. We needed to get the pressure up and neutralise the lactic acid in his blood with sodium bicarbonate. The adrenaline shot the blood pressure up to acceptable levels and the heart rate soared up to 140 beats per minute. A fit man, he would bounce back now we had things under control.

To finish the job properly I needed him under the bright lights of the operating theatre, together with proper sterile drapes and accurate monitoring of his bloods and vital signs. By now it was 2 am and the theatre was ready, the hospital corridors long since deserted. We'd just wheel him along with his chest wide open, clamp in place and a drape over the top to keep the wound clean, then lift him onto the operating table.

Throwing off my gown and rubber gloves, I retrieved the bullet from the floor. Things like that had a habit of disappearing, becoming desirable if macabre souvenirs. But this projectile had great forensic importance and I wanted to give it to the police, who were accumulating in great numbers.

I walked ahead of the bizarre cortège to scrub up again in theatre, where the nurses were waiting with the operating lights switched on. Now I could see. I gently removed the clamp and was met with a gush of dark blue blood from the pulmonary artery. The chest wound edges were oozing bright red blood and air was blowing out of a lacerated bronchial tube, but otherwise there was no problem.

In order to get a better view of the damage I pulled on the airless lung. It was what you'd expect from a high-velocity round, as if a dog had chewed at the vital structures. My hopes of conserving the lung rapidly disappeared. It simply had to come out, the whole thing. We needed to make him safe, not attempt some heroic repair job. If he died his family would be devastated – and the gamekeeper, as the culprit would turn out to be, left facing a murder charge.

I encircled the pulmonary artery with a thick silk ligature and tied it off. No more dark blue bleeding. Two large veins enter the heart from the lung; I tied them both off as well, then cut through all three large blood vessels with scissors. This just left his injured bronchus blowing out blood and froth. I stapled it, chopped through the tube and lifted out the redundant lung. It missed the receptacle and fell to the floor. He'd be just fine with the one, and the right lung is larger than the left. We washed out the empty space with warm salt solution and the powerful antibiotic gentamicin, infection now his biggest risk as the bullet had sucked fragments of jacket and shirt into his chest.

I sat and wrote up my notes while the registrar and house officer stopped bleeding from the wound edges and sewed him up. Documentation is vital in criminal cases, even at three in the morning. Driving home through the dark lanes I saw a fox on the grass verge, then a deer in the headlights, eyes sparkling. I was relaxed and content, another battle won, my adrenaline dissipated.

Our patient recovered without complications. The bullet matched the gamekeeper's rifle. He was arrested, then released on bail, having avoided a murder or manslaughter charge by a whisker – just minutes. It was a unique case for sleepy Oxford, one for Inspector Morse.

*   *   *

Nothing sets off an adrenaline rush like a stab wound to the heart. I still remember the first one I had to deal with as a young man, way back in 1975. I was a casualty officer in the accident department of King's College Hospital in south London, right on the edge of the war zone that was Brixton, London's equivalent to Harlem in New York, where I'd encountered many stab wounds. Having cut my teeth on chests at the Brompton, I was in my 'invincible' phase, a coiled spring ready to launch into action.

First a tutorial to set the scene. After my brief internship in Harlem I knew that most cardiac stab-wound victims died at the scene of the incident or on the way to hospital. Those who arrive alive sit on the edge of a precipice. The stakes are high, but most can survive with appropriate treatment, which is immediate surgery.

Most assailants attack face to face and stab the front of the right ventricle. A few wounds involve both right and left ventricles. Stab wounds to the left ventricle usually enter from the flank or back – the 'domestic incident' route. The thin-walled right atrium is protected by the breastbone, while the left atrium lies further back in the chest. Only rarely are the atria involved in knife wounds.

Rule number one. If the knife – or occasionally screwdriver – is still in place, do not remove it, and if it's bobbing about with every beat the blade or shaft is very likely plugging a hole in the heart muscle. Such patients are usually suicide attempts, as assailants rarely leave their knives and fingerprints as evidence.

When a knife is withdrawn, blood under pressure sprays out into the fibrous pericardial sac, the confined space that houses the heart. If there's free escape of blood out of the pericardium into the expansive chest cavity this will most likely result in exsanguination – bleeding to death. When blood accumulates within the pericardial space because the entry wound is small, we call that cardiac tamponade. As blood compresses the heart itself, the patient's blood pressure falls until a balance is reached and the bleeding stops. The circulation is maintained with lower blood pressure. These patients tend to survive. They're brought in pale, cold and restless with a fast heart rate and distended neck veins, but are perfectly able to live for a short period of time as long as their blood pressure is kept down.

Rule number two. Those admitted fully conscious usually have cardiac tamponade, and many need immediate chest opening for resuscitation. Holes in the heart are not amenable to standard resuscitation techniques, because if a patient is given intravenous fluids they will bleed more, often terminally. So it's important to control the bleeding point first. Once the cardiac tamponade is relieved the patient may not need any fluid. I've operated on tamponade patients to whom so much fluid has been given that their poor hearts were fit to burst. Before sewing the wound I'd have to open it up and discard copious amounts of diluted blood into the sucker. Only then would the heart look sufficiently comfortable for me to stitch the laceration.

Some patients arrive still warm but showing no other signs of life. But emergency surgery should only be undertaken if their pupils react to light. With vigorous cardiac massage and adrenaline it's possible to restart any heart, whether the brain is dead or not. That's why it's important to scrutinise the pupils first. No coroner will allow a murder victim to be kept alive just to be an organ donor.

I was still a junior doctor at King's, not a heart surgeon, and at two in the morning the department was full of drug addicts, drunks, vagrants and the walking wounded. Not that we didn't care for them. We did. The nurses were saints but constantly needed protection. It was a volatile environment.

This particular patient had been dumped in the entrance hall by fellow gang members. His shirt was covered in blood, and he was deathly pale and already unconscious. The porters brought him through to a resuscitation bay and the sister in charge called the resuscitation team. He still had a faint pulse and his pupils reacted to light.

As the nurses removed his shirt I could see the stab wound directly over his heart, about 1 cm wide. Blood trickled from the edges of the wound but his heart wasn't pumping, and his jugular veins stood out like tree trunks in his skinny neck because of raised pressure within the pericardial sac. This was an obvious case of cardiac tamponade.

The anaesthetist had already inserted the endotracheal tube and was frantically ventilating the lungs, but we still needed a large-bore cannula in his jugular vein for transfusion. A nurse took over squeezing the gas bag while the anaesthetist did the deed. He couldn't miss. Dark blue blood shot out the end at high pressure.

In those days there were no consultants in Casualty at night and there were certainly no cardiac surgeons in the hospital. The sister knew I'd worked at the Brompton. She just looked at me and said, 'Open him up. I'll help you.'

My brain said, 'Oh shit,' but my mouth said, 'Let's get on with it then – it's now or never.'

The anaesthetist was a senior registrar, and he nodded in approval, well aware that the kid would die if we did nothing. External cardiac massage is futile when the heart is compressed and cannot fill. There wasn't even time to scrub up as he had no pulse or blood pressure. The assembled crowd rolled him over left side up while I pulled on a gown and gloves, Sister following suit. I stood behind him, Sister in front, and with my own adrenaline pumping I carved his chest open with a scalpel, then spread his ribs with a metal retractor stored ready for such an eventuality but rarely used.

There was no blood or air in the patient's chest cavity as the stiletto knife had gone directly through into the pericardium and right ventricle. All I could see was the tense, blue, bulging pericardial sac. I knew what I had to do, assuming I could stop my own perspiration from clouding my eyes and dripping into the incision.

I opened the stretched membrane with a scalpel, and blood and blood clot spewed out. His heart was still beating but empty, and the ventricles filled as the pericardium emptied. His blood pressure started to come up and the stab wound began to squirt blood out again, although this was no longer an issue.

I put my index finger over the gash and said, 'Transfuse him while I stitch the ventricle.'

'What stitch do you want?' Sister asked.

I didn't have any idea and just said, 'Give me anything you've got on a curved needle.'

The first needle was much too big, the next much too small, but the third was just right – a blue-coloured, braided suture that knotted well. Perfect. I swapped fingers with Sister, who'd never touched a heart before, and she was squirted with blood.

Now the tricky bit. I mounted the curved needle on the needle holder and edged into the best position to throw the stitch. I knew that as soon as Sister moved her finger blood would be pissing out. Not only that. The young heart was now bounding away, a rapidly moving target and not easy to stitch accurately. Deep breath. Just get on and do it.

I drove the needle straight through the middle of the laceration from one side to the other, with deep bites. Sister cut the needle from the suture material and, to avoid tearing through the muscle and making the hole bigger, I tied the knot very gently. It worked, but to make him safe I needed further stitches on either side – three in total. It was nerve-racking for an amateur, as each time the needle penetrated the muscle it triggered a run of fast, uncontrolled rhythm. I guess in all the three stitches took me ten minutes, very different from now.

Sister looked straight at me over her mask. I knew what the eyes were saying. She was impressed. Actually so was I – the hero of the moment. The patient's blood pressure and heart rate were soon back to normal, and just when we no longer needed him the cardiothoracic registrar had been called in. By now I was happy to leave him to it.

Sister and I retired to the coffee room, very sweaty but elated. With his chest closed up, they rolled the patient onto his back on the trolley.

There was blood everywhere – on the stretcher canvas, in his hair, soaking his clothes and drying in a pool on the floor, all testament to our struggle. They needed to get him to intensive care for a clean-up. By now there were scores of other patients in the department, all getting restless with the wait.

Then the lad suddenly woke up, uncontrollably agitated following his near-death experience. He sat bolt upright and began tugging at the drips. The jugular vein cannula in the neck disconnected. As he took a deep breath in, negative pressure in his chest sucked air into the circulation and he collapsed, pulseless again for a different reason. At the time no one realised why it had happened. They started external cardiac massage but couldn't revive him. My first solo heart operation had proved fatal, and I'd gone from hero to zero in the space of a few minutes. Shit and derision.

Suddenly the night had turned into a nightmare and paranoia set in. I was concerned that I'd be blamed for the death and be accused of recklessness, but I needn't have worried. Sister and the anaesthetist made the situation clear; that without my intervention he'd have been dead sooner. The case went to the coroner's court. The verdict? Unlawful killing. The cause of death? Air embolism after a cardiac stab wound.

Not only was the operation my début emergency thoracotomy, it was also my first encounter with this fatal complication – air reaching the vessels of the brain. Sadly, it would not be the last. I was destined to operate on many more cardiac stab wounds throughout my career. Most were simple; a few were complex, involving the heart valves or coronary arteries. But none of the patients died.

Knives and bullets are not the only source of penetrating chest wounds. Some of the most horrific injuries occur during road-traffic accidents.

One quiet Saturday afternoon in the autumn of 2005 I was waiting for my son's rugby match to start when my mobile suddenly rang and I was required to scramble again. It was an immediately life-threatening injury in a young woman. Mark's school was just ten minutes from the hospital, and I was waiting in the hospital before the unfortunate victim arrived.

Input from the paramedics en route suggested that a car had veered off the A40 dual carriageway at speed, shattering a wooden fence. A sharp shard of wood the length of a spear had penetrated the windscreen and transfixed the driver's neck. The fire brigade had extricated her from the wreck, but she was blowing air from the wound and had difficulty in breathing. Her blood pressure was low, so they suspected internal bleeding.

While I waited with the trauma team in the resuscitation area warning lights were flashing in my brain – it sounded as if her windpipe had been cut in two. If this were the case, then blind attempts to pass the breathing tube through could push the ends apart and completely obstruct her airway. So I wanted an experienced cardio-thoracic anaesthetist to join us and the cardiac operating theatre team to stand by.

I called Dr Mike Sinclair myself, asking him to come at top speed, which he did. As we waited I politely requested that the resuscitation team should hold fire until I had the chance to examine the woman. It was already more than an hour since the crash, and if she was still alive it meant that she'd reached some degree of equilibrium. A couple of minutes to work out the likely injuries would be time well spent.

Tension rose palpably as the woman was wheeled in. She was awake but deathly pale, rigid with fear and her lips blue. All eyes were immediately drawn to the gash in the root of her neck on the right side, where bare sterno-mastoid muscle was exposed while air raised the torn skin as she exhaled. It sounded as if she were farting through the wound with every breath, while simultaneously spraying out an aerosol of blood. I was in no doubt about the cause. Equally, I was incredulous that the transfixion hadn't ripped out at least one of her two carotid arteries. If it had she'd have died at the scene.

The woman feebly raised her right arm, inviting me to take her sweaty hand. I was pleased to do that. We needed to connect as we'd be spending the afternoon together. Instinctively, I told her she'd be fine – not that I knew that, but she could use some comfort, to be treated like a person not an object of curiosity.

She was in shock, not just mental distress, and had clearly lost litres of blood internally. My guess was that the stake had passed downwards through her neck and into the left side of her chest, taking out a significant blood vessel. A good old-fashioned stethoscope would tell me that. Physical examination was quick and still important in this era of fancy scans. Air was filling her right lung but there was no sound of breathing on the left. When I tapped her ribs the left chest was 'dull to percussion', a traditional physical sign of fluid surrounding the lung. So she had blood in the chest and barely recordable blood pressure, with her heart rate 110 beats per minute.

Now we faced a stern surgical test – a severe injury to the root of the neck together with bleeding into the left chest. A tricky combination. Yet the basic principles remained the same. First establish a safe and reliable airway. Next take control of the breathing. Then support the circulation, in this case by stopping the bleeding and blood transfusion. The 'ABC' of resuscitation.

I needed Mike to put her to sleep. The only reliable way to secure her airway was with a rigid bronchoscope – a long, narrow brass tube with a light at the end. We had done hundreds of bronchoscopies together, whether to investigate lung cancer or remove inhaled peanuts from children.

By now the resus team had put a couple of drips in the woman's arms and were giving her clear fluids. I didn't want too much of that. She was critical but stable, the same old story. The blood pressure drops and a blood clot plugs the hole – nature's own rescue strategy. Clear fluid pushes the pressure up and makes the patient bleed again. 'Treating the numbers,' I call it, 'not the patient'. Then

Mike walked in and we agreed to push her directly round to the operating theatres. There I had complete control surrounded by my own team, away from the circus.

Sister Linda had the bronchoscope waiting in the anaesthetic room, but first Mike needed to anaesthetise and paralyse her. Then I could slip the tube down the back of her throat, through her vocal cords and into the injured trachea, just like sword swallowing but down the windpipe. High-pressure ventilation through the scope sprayed blood out of her neck and there was blood everywhere, but I could soon see the injury. Two-thirds of the circumference of her trachea had been lacerated, leaving only the muscular back wall intact.

I pushed a long gum-elastic probe down the bronchoscope and through the site of injury. After vigorously blowing in air to raise her oxygen levels the bronchoscope was withdrawn. Mike could then railroad his breathing tube safely over this guide. 'A' and 'B' were sorted. We could ventilate the lungs safely.

Now I had to get on with 'C' and stop the life-threatening bleeding. They wheeled the woman through into the operating theatre and turned her left side up. Dawn was already scrubbed up, with the thoracotomy instruments laid out on sterile linen. I didn't have to say a word. It all just happened around me like clockwork. Mike was ready with two units of donor blood and now had arterial blood pressure monitoring on the screen via a cannula in the wrist.

A range of thoughts went through my mind at the scrub sink. First, I was relieved for the poor woman that she was unconscious and far distant from her terrifying ordeal.

Then I was apprehensive. What would I find in the apex of the chest? I feared laceration of the large subclavian artery to her arm, although she still had a pulse at the left wrist. Hopefully it was just low-pressure venous bleeding, which would be easier to control. I was cognizant of the fact that the nerves to the arm were close by and I needed to avoid damaging them with the electrocautery.

Two litres of blood spilled out of her chest, splashing over my trousers and clogs and onto the floor. Warm and wet, but wasted. It was meant to be good for the garden. With compression relieved, her left lung expanded like a kid's balloon. It was virginal pink, not like the mottled grey lungs of smokers. We scooped and sucked blood from the depths of her chest until the ragged hole came into view. Mercifully there was no brisk, bright red arterial haemorrhage, just dark red bleeding from the main arm vein. I set about stopping the bleeding. If I tied off the vein her arm would swell, so I repaired it with a patch from a less important vein to preserve the flow.

Now I was content that she was safe, we washed the chest cavity out with antiseptic solution. All the other main arteries and nerves were clearly visible in the roof of her chest. The stake had simply pushed them aside, generously limiting its destruction to the least important structures. The luck of the woman was barely believable. 'C' for circulation was now sorted.

We were left with one other major injury to sort out – the transected trachea, a big tube containing air and much less intimidating than what we'd so far done. We closed up her chest, leaving a drain to remove residual air and blood, and I injected a generous volume of long-acting local anaesthetic into the nerves under the ribs to dull the pain. She had suffered enough.

It was time for a cup of tea while they rolled her onto her back, ready to explore the neck wound. I liked to operate in the neck. Hers was slender with no fat, making everything easier. The horrendous gash, just above the joint between the sternum and clavicle, was 8 cm long. It gaped widely to reveal bare muscle, like grinning lips exposing teeth. The simplest approach was to excise the ragged edges then extend the incision into a thyroid gland incision.

Her lacerated trachea was right in front of me, with the thyroid gland above and the rigid plastic breathing tube passing through the gap. With the benefit of full resuscitation the wound edges oozed bright red blood. This was easy to stop, but because the rural fence post was inevitably covered with bacteria I excised the contaminated edges of the trachea, then joined the clean ends with multiple separate sutures.

It had been an intimidating problem but easy surgery. I managed a solid, airtight repair and finished by checking the nerves to the vocal cords. Again these were spared what might have been. God must have been with her in the car. Or sitting on my shoulder. Maybe both. Mike gave her a slug of heavy-duty antibiotics for good measure, then we closed the skin and subcutaneous layers with metal clips. A job well done.

The family were huddled anxiously in the intensive care unit. They'd come in through the accident department, been injected with a dose of pessimism then dispatched for the long wait. Waiting to be told the outcome of emergency surgery is a truly miserable experience, particularly when it's your own kid and they tell you that a fence post has nearly taken her head off. Alive or dead? Disabled or intact? Disfigured or still beautiful? It's difficult to concentrate on the football results.

I told them what I'd told her when I squeezed her hand as life ebbed away – that everything would be OK. Then I rode off into the sunset. Well, as far as the pub, that is, for time with my own little family, to hear about my son's rugby and my daughter's golf match. The fights. The cuts and the bruises. And that was just the ladies' golf.

As for the woman, she recovered quickly. Mike and I came in on Sunday morning to find her wide awake, so we took the bull by the horns and bravely pulled out the tracheal tube. Inevitably, after such an accident she felt like she'd been hit by a truck. Her throat and chest were sore, but her breathing was fine and she could speak. Everything was intact and she went home in a week.

Thankfully, with age, my adrenaline addiction and testosterone are wearing off, yet the excitement of the unexpected persists. For the unfortunate patient, any prospect of survival depends upon having an experienced trauma surgeon at hand. Few are offered that privilege.

14

despair

Strength does not come from winning. Your struggles develop your strengths. When you go through hardships and decide not to surrender, that is strength.

Arnold Schwarzenegger

OXFORD BROOKES UNIVERSITY lies within a mile of my hospital and is full of vibrant, happy students. One of these, a twenty-year-old girl studying Japanese, had complained of fainting attacks. A series of preliminary investigations, including an ECG and echocardiogram, indicated that her heart was normal, but one evening she was talking with friends on the campus when she suddenly collapsed to the floor.

This happened only a few days after the very public resuscitation of a Premier League footballer on the pitch in north London, which had been widely reported in the media. He survived through effective bystander resuscitation by a cardiologist in the ground, then rapid transport to a frontline cardiac centre. As a result, cardiopulmonary resuscitation was very much in the public awareness.

The girl's friends began cardiac massage and called the emergency services. A paramedic ambulance was despatched from the nearby headquarters and reached her in less than four minutes. Their cardiac monitor showed ventricular fibrillation – random electrical activity, with the heart squirming aimlessly and not pumping. These days, paramedic vehicles carry a defibrillator. As the girls continued chest compressions the paramedics set up to defibrillate, putting electrodes on the front and side of her chest. Ninety joules. Zap!

This usually works in heart attack patients, but after a brief period of flatline it fibrillated again. Although the hospital was only two minutes away from the campus, and full of specialist doctors, they didn't bring her in. Instead they inserted a tube into her windpipe and persisted with on-site resuscitation. At least she was getting oxygen. The ambulance was carrying a new toy – a 'Lucas' chest compression machine. Manual cardiac massage is tiring but this machine doesn't tire, rhythmically pushing down on the lower half of the breastbone, forcing blood out of the heart and around the body.

When more shocks failed they fitted the machine around her chest. Now her heart was squeezed between the sternum and spine and continuously pounded like a meat tenderiser. Time passed, and it was more than thirty minutes after the cardiac arrest that she was wheeled in to the accident department, lifeless but replete with medical equipment, the Lucas machine still bashing away. Her pupils still reacted to light – they'd kept her brain alive – yet her poor heart was still squirming, battered and bruised.

Fabrice Muamba, the Bolton Wanderers footballer, had been lucky as there was an experienced cardiologist right there in the ground. What this young woman needed was treatment targeted towards the underlying problem, but what she got was the Standard Advanced Life Support approach: first, defibrillation with high electrical charge – 150 joules then 200 joules on multiple occasions using the original electrodes – then, after repeated failure with this and persistent fibrillation, continuous cardiac compression by the machine and adrenaline injected into a vein. The adrenaline would have been potentially useful had the heart been contracting, but it worsens muscle irritability and predisposes the patient to more ventricular fibrillation.

As a further measure she was given the drug amiodarone in an attempt to calm the electrical storm – a good move, but after thirty shocks she still reverted back to ventricular fibrillation. With the situation looking desperate, the on-call cardiologist, Dr Bashir, arrived. He looked carefully at the patient and changed one thing: the position of the electrodes on the chest. He placed one on the front of her chest, over the right ventricle, and one on her back, directly behind the left ventricle.

One shock of 200 joules, and normal heart rhythm was restored. With adrenaline on board, her blood pressure immediately rose to above-normal levels, but although this had the benefit of increasing blood flow to the traumatised heart muscle it also increased electrical instability. The result? Repeated fibrillation that needed more shocks and a high dose of a beta-blocker to counter the stimulant. Once the electrodes were in the correct position the shocks worked each time. Dr Bashir, an experienced electrophysiologist, then prescribed a combination of powerful rhythm-stabilising drugs in high doses.

Around two hours after she'd collapsed the girl's erratic rhythm began to settle and she became stable enough to perform an echo study to obtain pictures of the heart. Whatever this showed would be important. Only a handful of problems cause sudden death in young people. One of the possibilities is an inherited condition of thick heart muscle, but the echo soon ruled that out as both ventricles were normal in size and thickness.

By now her right ventricle was suffering visibly after the prolonged cardiac massage and electric shocks. It was dilated and contracting poorly, although the heart valves all appeared normal. Very rare coronary artery abnormalities can cause ventricular fibrillation, but from what could be seen of these small vessels they also appeared normal.

Was she suffering from a primary ventricular dysrhythmia, electrical instability in a structurally normal – albeit now battered – heart? This can present as fainting attacks or sudden cardiac death without any identifiable genetic syndrome. It's not related to exercise or stress but probably arises from within the heart's own electrical system, and it can take place in short bursts of electrical instability or as a full-blown 'electrical storm'.

If it settled it could be treated by electrical mapping to locate its origin, followed by destruction of the irritable source. This was Dr Bashir's specialty and would be performed in the catheterisation laboratory. It can be done during the electrical storm if the patient's circulation can be sustained in the meantime, but this is not easy to organise at night as it needs a highly trained support team.

The plan was to move her from the accident department to the cardiac intensive care unit. The intensive care consultants were already involved, working to normalise her blood chemistry after three hours of resuscitation. They were anxious that she was sliding into heart failure and wanted my opinion as to whether she'd need mechanical circulatory support.

I arrived at the accident department at 9.30 pm to find a crowd around her bed in the resuscitation area, most of them spectators doing nothing. The cardiac massage machine was still in place but had thankfully been switched off while her rhythm was normal. Personally, I disliked it. Cardiac massage has its place, but the heart is a delicate organ, and one I don't like to see being pulped by a machine. By now the intensive care doctors had her well sedated and ventilated, and her blood chemistry had improved because normal rhythm had given her much better blood flow. The cardiology registrar lingered nervously by the defibrillator.

I'd only been there three minutes when she fibrillated again. This time no pounding on the chest was administered, just a finger on the trigger of the defibrillator. Zap!

Her heart returned to its normal sinus rhythm. I suggested that we take her round to the cardiac intensive care unit, away from the circus, and get the sledgehammer back into the ambulance away from her broken ribs.

After seventy electrical shocks we settled on that diagnosis of idiopathic ventricular fibrillation. By this stage she was beginning to respond to the anti-dysrhythmic drugs, so perhaps it was wise not to move her to the catheterisation laboratory while we appeared to be winning. As the shocks became less frequent, the heart was easier to defibrillate.

We stayed with her in intensive care, right by her bedside. During the night her parents and boyfriend arrived after a terrible journey down from the north of England, all pole-axed by grief and anxiety. This was the worst part for me. I was watching as the nurses told them the story away from the bed, then witnessed the shock on their faces when they first saw her – on the ventilator, pale with blue lips, big drips in her neck, arms and wrist. This is how intensive care always looks, but it comes as a terrible shock the first time, and even more terrible when it's your child hovering between life and death.

Then I heard the quiet descent into recrimination. How could this have happened? She seemed so happy at Brookes. Did she get the condition from us? This was the time for me to ask her parents questions, but I asked my registrar to do it as I just couldn't face it. I hovered in the background. Had anyone in the family died suddenly? Was there any history of heart disease? Had she had problems before? Each drew a blank.

I knew what to expect next, which is why I stayed, although I hoped that it wouldn't happen. When the adrenaline wore off, the electrical irritability lessened but her blood pressure started to creep down, and by the early morning it was worryingly low. Meanwhile the pressure in her veins drifted upwards as the right ventricle – battered and bruised, now buggered – struggled to cope. The urine flow tailed off, as it always does in these circumstances, and the acid in the blood started to rise, produced by the muscles as their blood flow dwindled.

She needed more shocks, and unfortunately there was no time to usher her parents away. This was a grim reminder to them that she was actually dying. Her hands and legs were cold with the onset of cardiogenic shock that hadn't come from the dysrhythmia but from the heavy-duty cardiac massage and repeated electric shocks, and certainly not helped by the high-dose beta-blocker she'd needed to counteract the adrenaline.

I asked for another echo, this time taken through a probe in the gullet. The camera sits right behind the heart so the pictures are much better. Things had changed dramatically for the worse, as both the left and right ventricles were contracting poorly. This is when the 'what ifs' kick in. Would this have happened if the defibrillating electrodes had been positioned differently? What if they'd just brought her straight into the hospital so she could be treated sooner by those who could make the diagnosis, then target the drug treatment as my colleague had done? What she'd needed was expertise and drugs, not a mechanical sledgehammer out in the town.

'What ifs' are no good in cardiac surgery – they simply don't help. We just have to get on and treat what lies in front of us. I knew what she required now. Her struggling heart was still recoverable but she needed circulatory support, and the only thing we could do quickly was to insert an intra-aortic balloon pump, in full awareness of the fact that it was pretty useless in shocked patients. It went in anyway and slightly improved the blood pressure on the monitor. But she needed more blood flow and the balloon doesn't provide that. We had to give her the vaso-pressor drug noradrenaline to keep the pressure above 70 mm Hg, but that soon triggered further episodes of ventricular fibrillation.

What I meant when I said she needed circulatory support was a ventricular assist device to take over the circulation, the sort of pumps we had before the money ran out. In this case we needed a system called extracor-poreal membrane oxygenation – or ECMO for short. This combines a centrifugal blood pump with an oxygenator and is similar to the oxygenator in the heart–lung machine, except that it's engineered for long-term use and is safe for days or weeks until the heart gets better. We needed this because she had both left and right ventricular failure and her lungs were deteriorating in response to the shock. But we didn't have one. Only a handful of UK units were funded to use it, primarily for young patients with severe lung disease.

By now my own blood was beginning to boil as I saw the despairing parents by the bedside, as I watched the watery spring sunshine break over the horizon, as normal, healthy people were beginning their day, just as she did yesterday at Brookes.

So what did the recent NICE guidelines for acute heart failure have to say? They said that one should 'ask advice from a hospital that has circulatory support equipment'. We did. My surgical colleagues who I'd trained said she needed ECMO. But what were the prospects for safe transfer of a dying girl who was suffering ventricular fibrillation at regular intervals? Who'd been shocked seventy times? Whose heart was toast? The odds of getting her to another centre in safety were negligible. No one disputed that fact.

Given our track record of innovation, my colleagues expressed surprise that we had no ECMO system and said that I simply had to get the equipment brought to Oxford by the company representative as soon as possible. We couldn't locate the supplier until 8.30 am, by which time her blood pressure had sagged again, with a rise in pressure in the veins. As a result her tissues were poorly perfused, blood flow within the vital organs was critically impaired and acid levels started to climb.

I debated whether to take her to theatre and put her on a conventional heart–lung machine. Yet this could have been a disaster for a number of reasons. It would further damage the lungs and the ability of the blood to clot. Bleeding is the commonest life-threatening complication during ECMO, and after prolonged standard cardiopulmonary bypass the risks would be even greater.

But there was one other option that would buy us some time, a powerful heart failure drug called Levosimendan that we'd used in the past. It helps link calcium to the muscle molecules and makes the force of contraction greater, doing so without increasing tissue oxygen uptake or ventricular irritability. I asked the intensive care doctors to start an infusion of the drug, only to be told that we didn't keep it any more as the hospital said it was too expensive. All we had were drugs to constrict the blood vessels and make the heart more irritable, or drugs that would flog the heart and make things worse.

The ugly truth was that we were desperately trying to keep this young woman alive without the equipment or the drugs she needed. It was a tense, miserable morning watching the clock tick away, while attempting to reassure the poor parents that we were doing everything we could. We waited for the ECMO equipment to arrive, in the meantime infusing vials of sodium bicarbonate to neutralise the acid and watching her pupils. Did they still react to light? Was her brain getting enough oxygen? Higher doses of arterial constrictor drugs would briefly elevate her blood pressure, hopefully increasing flow to her brain – but all at the expense of her limbs and gut. Her hands and legs were already cold and white, and she had critically low blood flow, with acid pouring into the circulation from muscles starved of oxygen.

By midday I couldn't watch any longer. I went round to the operating theatres and told them that we had to put her on cardiopulmonary bypass, hopefully just for a short time until the ECMO equipment reached the hospital. Then someone asked the inevitable. Who was going to pay for ECMO? Who was going to look after it at night? What if?

I was tired and irritable so I let rip. Who on earth were they to question our efforts to save a twenty-year-old? So bloody what if we weren't a transplant centre. She didn't need a transplant. Her own heart just needed a rest from the battering it had received over the last twenty-four hours. Why was this so-called 'centre of excellence' unable to save a kid who'd collapsed within a mile of the hospital? It certainly wasn't through lack of effort by the medical staff.

Just as I was about to lose it altogether I heard that the equipment had arrived. Our patient was already on her way round to theatre, so I went to meet the company representative who'd made an enormous effort to come and help us. He'd already been in Oxford for more than an hour, stuck in traffic while trying to get into the hospital, then driving round in circles to find a parking space – all with mounting levels of frustration and anxiety. Lost time, lower chance of survival. He knew it and was extremely pissed off.

Once the equipment was ready it took just minutes to establish the ECMO circuit via blood vessels in each groin. Ultrasound visualisation showed the femoral artery to be narrow, and because of this I chose to surgically expose it and join a vascular graft to its side. This would guarantee that the leg still received adequate blood flow. The femoral vein in the opposite groin was cannulated directly using a needle and guidewire. The long cannula was advanced into her right atrium and positioned carefully using an echo probe in her oesophagus.

When the pump was turned on her blood pressure immediately rose to 110/70 mm Hg, while the pressure in her veins fell from 25 mm Hg to 5 mm Hg. Although we'd introduced a kidney dialysis cannula in her neck, her urine flow improved in response to increased blood flow. She'd been transformed by the ECMO system within a few minutes – better colour, better chemistry, different patient. I was jubilant, and her parents finally relaxed.

For the first few hours her pupils remained responsive to light. Then late in the afternoon, when her heart had improved substantially, her pupils suddenly dilated widely and became unreactive to light. My complete nightmare scenario – body better, brain buggered. Starved of blood and oxygen, her brain had started to swell. The pressure within the bony confines of her skull went up and her brain stem herniated into the spinal canal. Medical jargon for fucking disaster.

At the time I was lying on the sofa in my office, hoping that the battle was at last over. Sue, my secretary, knocked tentatively on the door before she headed for home. The intensive care unit were asking that I go back, a message that always gives me that sinking feeling. No one calls with good news – it's always trouble. I expected bleeding, or something that I could sort out. But when I reached the bed space the curtains were completely pulled around.

Her parents were sitting on either side of her, each holding a hand, and now completely exhausted both physically and emotionally. I needed to know what the issues were before I disturbed them. The distraught nurse looking after her came out to talk. Her pupils had blown quite quickly, and I needed to know the cause straight away – whether she had intra-cerebral bleeding from the heparin anticoagulation or a swollen brain through lack of oxygen.

A brain surgeon might be able to help with the first, by removing the blood clot. The second would more than likely signal a fatal conclusion to our efforts, just when we'd beaten off the ventricular fibrillation. Four hours had passed since the last shock, and now we needed to go to the brain scanner as soon as possible. I went to arrange it myself, then asked a brain surgeon colleague to come and look at the results with me.

Brian scans emerge slice by slice, showing multiple cross-sections through convoluted grey and white matter. It's complex but well-charted anatomy, each part responsible for part of our lives, some parts more important than others. The skull is a rigid box, so when the brain swells something has to give. The fluid spaces are compressed and disappear, the delicate brain appendages and nerves are distorted, and eventually parts of the brain stem are pushed out of the skull, hence the loss of the pupils' reaction to light. And when brain stem reflexes are lost, the patient is dead.

The whole scan was finished in minutes, then the slices were computed into a three-dimensional reconstruction of the whole organ. It told a story I didn't want to hear. 'Severe brain swelling with herniation of brain stem through the foramen magnum,' was the official radiologist's report. I tried to persuade the brain surgeons to take the top of the skull off to decompress the brain. They were sympathetic but said it was too late. Sorry about that. But not as sorry as I was.

We pushed her back to the intensive care unit. This was quite an undertaking in itself with all the equipment – ECMO circuit, ventilator, balloon pump, monitoring equipment – and we moved slowly now in a sad procession.

What were we left with? All her other organs were recovering. She was warm and pink, flooded with well-oxygenated blood from the machine, kidneys making urine, gut absorbing food and liver removing toxins. All organs need blood and oxygen, and ECMO – this simple, inexpensive technology – supplied both in abundance. But it had been too late for the brain. The cells that we'd failed to save were the cells that mattered the most.

I was very bitter about this. No other team in the UK had our breadth of experience, had put in the same amount of hard graft in the laboratory, had made the important discoveries. But that didn't matter. What mattered was that we weren't a transplant centre, so we weren't eligible for funding. What mattered was keeping down costs. Death comes cheap.

I couldn't face telling her parents. I took the coward's way out and went back to my office, black as thunder. The intensive care doctors did their best to treat the brain swelling with drugs but they were just going through the motions. The die had already been cast. ECMO was withdrawn after forty-eight hours because of the brain death issue. I took the tubes out myself. By now her heart was working well – good blood pressure, normal rhythm, no ventricular fibrillation. That battle had been won.

After formal tests for brain stem death, the issue of organ donation was raised with the desolate parents. Apparently the girl had previously expressed the wish to donate her organs in the event of her premature death, and her parents were in accord with her wishes. Before this I went to see her while they were still there. The nurse who'd helped us battle for her life was also at the bedside as she wanted to stay to the end, to see it through and support Mum and Dad. Uncommon decency. That takes moral fibre and courage.

What could I say at this point? I was really sad. My son was a fellow Brookes student of similar age. How would I have felt in their shoes? I didn't have to think about that – I'd faced so many bereaved parents that I already knew. What I told them was this. I was really sorry for their loss. Given the difficult circumstances, an experienced team of consultants had fought day and night to turn things around. All of my colleagues were devastated by the outcome and we appreciated the kind offer to donate her organs. It was a gesture that would transform the lives of others.

In the end she donated her liver and both kidneys, so three patients benefited. The fact that these organs still functioned normally was a testament to ECMO.

Within days we needed the equipment again, this time for a young woman who'd just given birth and had suffered an embolism of uterine fluid to her lungs. All I could advise was to send her directly to an ECMO centre, knowing full well that the delay would prove fatal. Unfortunately, I was right.

Then I could have used it for a forty-year-old patient who'd suffered an accidental air embolism and cardiac arrest in our own intensive care unit. She died. And so the saga goes on.

The young woman's death caused real distress amongst her university friends and teachers at Brookes, so much so that I wrote to the vice-chancellor, expressing my own regret that we couldn't have saved her after her friends had made such a valiant effort when she collapsed. Months later I received an invitation to the university graduation ceremony. They intended to award her a post-humous degree, and would I please come with the parents?

I sat with her mum, dad and boyfriend in the front row, and we watched the bright young men and women come up to the stage to collect their awards. Then the chancel-lor Shami Chakrabarti gave an explanation about the special award and thanked the surgeon for his valiant attempts to save her. Someone had to go up and receive the certificate. Mum was the one. Dad was frozen with grief and her boyfriend sat desolate. I was choked. I couldn't speak, but I helped the poor lady stumble up the stairs. This was not how it was meant to be, not the antic-ipated end to her university career. All her friends and tutors rallied round. The family were happy to see them and bravely they went to the reception.

But I was bitter and twisted. I went away crushed, the weight of the world on my shoulders. It was the saddest day of my whole career.

In memory of Alice Hunter, so others might be saved.

15

double jeopardy

When I was young and full of life,
I loved the local doctor's wife,
And ate an apple every day
To keep the doctor far away.

Thomas W. Lamont,
My Boyhood in a Parsonage

JULIA WAS FORTY. Pretty, blonde and feisty, she had a busy career in London. At the weekend she was an accomplished event rider, not far from the top tier, often rubbing shoulders – or bridle – with the best. So she'd left it late to have her first baby. But it would be fine as she was fit, both physically and mentally. Indeed as a psychology student at Durham she'd played hockey for the university, then for her county Leicestershire. And football. And cricket.

There was one funny thing, however. She could never complete a bleep test, something always holding her back. And she regularly fell asleep in meetings, so much so that she was admitted to a private hospital for sleep studies. They suspected narcolepsy, but nothing was found and it cost a fair bit.

She was deliriously excited when the pregnancy test turned blue in April 2015 in only the second month of trying. Bingo! But then she started to get really tired, then a bit breathless, then more breathless – just getting onto the horse – although she was reassured that it was normal in pregnancy, and was all down to hormones and retained fluid.

Desperate not to let her exhaustion beat her she started running again, determined to get fitter. The first time she pushed herself for five kilometres, but the following week she was breathless by the end of the street, with a burning throat and a tight chest. Her breasts were tender and swollen, and she thought sore breasts could be part of the problem. She just had to slow down a bit, but at least she could still ride.

At thirteen weeks she saw the midwife at the doctor's surgery on a Monday. She was advised to take aspirin as prophylaxis against pre-eclampsia – the dangerously high blood pressure some women suffer during the later stages of pregnancy. She mentioned how unfit she felt and how quickly the situation had deteriorated. Instead of dismissing Julia as neurotic, the midwife's response was that she should get her heart and lungs checked out, promising to have a word with her doctor. Good for the midwife – it was a critically important decision.

The doctor listened as doctors should. He was kind and reassuring. 'Blood volume increases by a third during pregnancy,' he told her. 'It can make you breathless. Let me just listen to your chest.' Then his tone changed and he suddenly looked serious. 'Just a slight murmur. But we should get someone to take a look at you quickly.'

Soon he was on the phone to the Windsor Clinic in Maidenhead. A cardiologist would see her on Wednesday, the day after tomorrow. Julia was anxious but went back to work. United Biscuits needed her and it would keep her mind off the 'murmur' word.

The Windsor Clinic had a nice waiting room, an efficient receptionist and a comfortable sofa, although none of this mattered a bugger to Julia. There were two important tests planned before she saw the cardiologist. So off with her smart black dress and on with the ubiquitous white gown; it showed her bum as it had ties down the back that you just couldn't reach.

First the electrocardiogram. She got up on the couch and a lady asked her to pop the top of her gown off. Sensors detected the electrical activity on her wrists, ankles and across her chest wall, then the ECG machine rapidly spewed out a long strip of pink paper with a black squiggly line. This was very important to doctors, completely meaningless to anyone else, but the technician said it was fine. How reassuring was that! Except it wasn't fine.

To the trained eye Julia's ECG showed what we call left ventricular hypertrophy – heart muscle under strain. Next she had an echocardiogram, the non-invasive window on the heart, which took ultrasound pictures with a probe and projected them on a screen. This time Julia coloured up a bit as it was done by a man, although he was nice and chatty as he smeared viscous jelly on her chest. All part of the job.

It took a while to get good pictures. He worked around her swollen left breast, trying not to hurt her. He started with her heart chambers – the left and right ventricles – seen best through the 'four chamber' view. The left ventricle was thicker than expected. The right ventricle, left atrium and right atrium all normal. But the ultrasound hadn't yet got the money shot. He shifted the probe to the top of the breastbone and angled it downwards.

Then his demeanour and expression changed abruptly. He went quiet and fiddled with the probe, and Julia sensed impending bad news. Her heart sank and she experienced that sudden cold, empty feeling, like your guts just fell out.

'What is it?' she couldn't help asking.

'Tight aortic stenosis.' It sounded like an automated reply. 'I'm so sorry. I'll go and tell the doctor.'

Then another lady came with a different echo machine to look for the baby, this time smearing slimy jelly on her belly. It was Julia's first introduction to her foetus, and there was concern whether it was still alive. Then, from the ensuing dialogue, she sensed that it might be better if it wasn't. While Julia's day was unravelling, the foetal heart was still beating away normally at around 150 beats per minute.

It was now time to see the doctor, a smart young cardiologist who also worked in the NHS. He had already reviewed the investigations, and while he knew the diagnosis there was nothing he could do to help. But at least

Julia had her clothes back on now, so she felt less exposed, less physically vulnerable, although she was verging on psychological meltdown. From her undergraduate studies she knew a lot about psychology, yet it didn't make it easier to control her own.

She spoke first, with no exchange of pleasantries. 'I'm in trouble, aren't I?'

'Yes, I'm sorry.'

That bloody word again. All doctors used it but none of them meant it.

'You've got very severe aortic stenosis. Congenital aortic stenosis, in fact. Didn't anyone hear the murmur before you decided to have the child?'

Julia thought carefully. Other doctors *had* listened to her chest, yes; but no, no one had mentioned a heart murmur.

When the valve becomes very tight it can be difficult to hear one. Now it was very tight and her symptoms had been unveiled by the expanded blood volume – the extra work that a heart has to do to support the placenta.

To explain the physiology of what had happened, it's important to appreciate that from the twelfth to the thirty-sixth week of gestation the volume of blood pumped by the heart rises to a peak of 50 per cent above non-pregnant levels. Julia had hit the buffers by thirteen weeks because she had a severe narrowing of the valve at the outlet of the left ventricle. The crushing chest pain when she exercised stemmed from poor flow in the coronary arteries. When the pressure in her arm was 100 mm Hg, it was 250 mm Hg in the left ventricle – dangerously high. In addition, blood trying to enter her heart was held back in the lungs, causing them to be stiff. Further strain of any kind could cause oedema fluid to flood the lungs, risking sudden death. And Julia thought she was fit!

Now the *coup de grâce*. The life expectancy for severe aortic stenosis without the baby would be between a further six and twenty-four months at best. In her current situation it was only weeks. It was far too dangerous to continue with the pregnancy, so the cardiologist felt that he should arrange an abortion to take place before the weekend. Then it would be possible for her to have an operation to replace the aortic valve. She needed it soon.

This was not at all what Julia wanted. She'd left it late to have kids, but after three months of excitement and expectation she was attached to her foetus. And not just by the placenta. What if she never had another chance? She felt fine as long as she was doing nothing. Surely she could just do nothing until the baby was born? Simple logic and a price worth paying. But wrong. The cardiologist was in no doubt whatsoever that if nothing was done both Julia and the baby would die long before it could be delivered, even if it was delivered prematurely in twenty weeks' time.

Her options were limited. No surgeon would operate on her aortic valve while she was pregnant. If she wanted he would discuss her case the following day at a multi-disciplinary team meeting with a group of cardiologists, surgeons, intensive care doctors – and in Julia's case, obstetricians – who would review the information in detail, consider the alternatives and recommend the right thing.

But Julia was no wilting violet. 'What about my opinion?' she insisted. 'I want to keep my baby. Not have
people ganging up on me. What's my best chance of keeping the baby?'

This wasn't an easy question to answer. There was no
straightforward solution. He thought for a minute, then
said, 'I'll get you to a cardiologist in Oxford who special-
ises in heart problems during pregnancy.'

The ethical principles pertaining to pregnancy are
straightforward. The doctor's first responsibility is to the
mother, and while a baby may be sacrificed *in utero* to
sustain the mother's health, it's not acceptable to put a
mother at risk for the sake of an unborn child. A baby will
normally survive if delivered after thirty weeks, even
twenty-eight weeks. But only rarely have dying mothers
been kept alive with the sole objective of sustaining a
foetus.

The cardiologists at the district general hospital saw
the echo images. They judged the valve to be far too tight
for Julia to reach thirty weeks' gestation for a Caesarean
section. The hormonal changes and the increase in blood
volume were already life-threatening and she wouldn't
survive another sixteen weeks. Everyone's opinion was
the same – Julia should be advised to have a termination
of pregnancy within days, then have the aortic valve
replaced soon afterwards. An abortion would turn a
complex problem into a simple one, assuming you judge
cardiac surgery to be simple.

The cardiologist phoned her at work that Thursday
afternoon and summarised the depressing consensus from

his colleagues. She winced at that 'sorry' word again, but he'd arranged an appointment for her to see Dr Oliver Ormerod in Oxford the following afternoon, on the NHS. He emphasised that there was no time to spare and that in the meantime she absolutely must not ride or do any exercise of any sort.

Getting to the appointment itself was a nightmare – traffic queuing on the main roads to reach the hospital, traffic queuing to get into the car parks, no parking spaces, no help. She was going to be late for the most important appointment of their two lives and on top of it all she had that crushing chest pain again, followed by crippling anxiety. Last Friday she was an excited mother to be. Now she was filled with impending doom.

Oliver changed all that, as he was altogether different. Not wearing a suit or tie, and not seeming to take things at all seriously, he reminded Julia of one of her childhood favourites, Popeye the Sailor Man. He made her feel that she was the special one in that consulting room.

'You want to keep your baby? Let's see how we can help you with that.'

The tightness in her chest disappeared, a wave of relaxation flowed through her body and her hand involuntarily dropped down to the little bump as if to say, 'Don't worry! This doctor will look after us both.'

So what were the possibilities for keeping Julia safe and the baby alive? Oliver agreed that the valve couldn't wait until the baby was viable at twenty-eight weeks' gestation. Therefore the valve would have to be dealt with while trying to preserve the pregnancy. There were two potential ways forward. The first was a balloon dilatation of the critically narrowed valve orifice as a temporising manoeuvre, to buy them some time. The second was to get on with open heart surgery on the heart–lung machine. All previous medical opinion had been against the latter.

Balloon dilatation was done in the catheterisation laboratory under X-ray guidance but the uterus could be shielded from the radiation. The balloon would be inflated within the narrowed valve orifice to split and open the fused parts. If this would carry Julia through to thirty weeks the valve could be replaced after the baby was delivered. She'd then face safer heart surgery as a new mother.

My colleague Professor Banning was an expert in balloon valve interventions, and Oliver needed more detailed echo pictures of the valve to show him. If he agreed, the procedure would be carried out early the following week. But what were the risks? The valve might split and leak badly, causing acute heart failure, so a surgical team would need to stand by in the operating theatres. Alternatively, the valve might not open sufficiently to make a difference. Either way there was a significant risk to the mother and baby. It was not straightforward.

Oliver decided to admit her to the cardiology ward after the weekend. In the meantime he would talk to the only surgeon he knew who'd operated on patients in similar circumstances.

Oliver called me at home on the Friday evening and we talked about our previous experience. The last pregnant patient we'd managed between us had an unusual murmur discovered at twenty-eight weeks. She was found to have a massive but benign tumour in the left atrium, a left atrial myxoma like Anna's. We watched her carefully for four weeks in hospital, then delivered the baby by Caesarean section in a cardiac operating theatre at thirty-two weeks. Three days later I removed the tumour. Both did well.

Before that we'd treated a young woman with an infected artificial heart valve that was disintegrating and leaking badly. We took her to the operating theatre at thirty-three weeks and performed the Caesarean section, then I re-replaced the aortic valve at the same sitting. Mother and baby did well, although we had problems with bleeding from the uterus.

Then I reminded Oliver that in another hospital I'd replaced an aortic valve in a thirty-five-year-old with a twenty-week foetus. The valve replacement was fine and the baby had a detectable heartbeat afterwards. But in the middle of the night she aborted and haemorrhaged profusely. We came close to losing the mother as well as the baby.

Heart surgery in pregnancy is one of those rare procedures in which you can actually lose two patients – mother and baby. I'd read and analysed every published report about heart surgery during pregnancy, then produced a detailed review. At the time there were only 133 cases worldwide, only nineteen being aortic valve replacements. No mothers had died but seven of the babies were lost. Was this reassuring? No.

The big problem is that surgeons prefer to report

successes, so there could have been hundreds of unreported cases where the babies – or even the mothers – had died. But best keep schtum about these ones, eh? It's human nature. Nevertheless, we had some statistics to share with Julia and her family.

Oliver then asked me what I thought about the balloon option. I said it was a good idea but there were practical issues. Most congenitally deformed aortic valves didn't have defined parts that would separate under balloon pressure, not like rheumatic mitral valves for which the technique was well established. It was essentially a blind procedure – the valve might be destroyed, and the aorta might even split and bleed torrentially. We needed to ask Banning what he thought about the chances of success. If they decided to take the valvuloplasty route, I'd do my best to provide back-up. We left it at that.

After the weekend Julia was brought back into hospital for more tests. News of her pregnancy dilemma had spread fast, so there was an impressive turnout at the congenital heart team meeting at the crack of dawn on Thursday morning. We were joined by our paediatric cardiology colleagues from Southampton, and Oliver presented the case with superb new pictures of her heart.

Julia's aortic valve orifice was a narrow slit, and instead of having three cusps there appeared to be just one, what we call a monocusp valve. This looked like a rocky volcano, and was almost 1 cm in depth and rigid. The muscle below was ominously thick and it was curious that she'd reached the age of forty in this state. Would the balloon make a difference? Unlikely. Was it safe? Unlikely.

Then the bottom line. They'd already made their decision. She should go directly for aortic valve replacement with a biological prosthetic valve, a valve that does not need anticoagulation, which would endanger the pregnancy. This is what Julia wanted. It had been her decision, and she disliked uncertainty. She wasn't just feisty, but brave too. No one in the meeting disagreed.

Would I do it? Well, it would have to be quick and she must spend the shortest possible time on the heart–lung machine. While cardiopulmonary bypass is perfectly safe for the mother, it often proves the trigger for foetal death as the uterus and placenta don't like it one bit. The clear fluid that's used to fill the bypass machine dilutes the mother's blood, and this dilutional effect drops the concentration of the pregnancy hormone progesterone, making the uterus less stable and increasing uterine irritability. When uterine contractions appear during cardiopulmonary bypass it's an important predictor of foetal death. Next, if the foetal heart rate decreases through decreased placental blood supply and low oxygen levels in the baby's bloodstream, it can trigger a distress response. This raises blood pressure and puts pressure on the developing foetal heart, which often fails to recover.

I explained how to manage cardiopulmonary bypass in a pregnant woman. We needed to use higher pressure and flow than normal, and avoid cooling so that the placental blood vessels wouldn't constrict. Rapid surgery was vital. The cardioplegia solution that was needed to protect Julia's thickened heart muscle contained high levels of potassium and the baby's heart would be very sensitive to elevated levels of this. Too much maternal cardioplegia could stop it.

So we had to monitor the foetal heart rate and uterine contractions, and if we detected contractions we could infuse the pregnancy hormone progesterone to dampen them down. We could also increase the heart–lung machine flow if the foetal heart rate dropped. As long as everyone knew precisely what to expect I felt we stood a good chance of keeping the baby alive.

By now the mood had shifted away from terminating the pregnancy towards keeping the little family together. But we needed back-up. Should the baby die and spontaneously abort during the night the gynaecologists needed to be prepared, as they might have to treat uterine bleeding in a patient who'd just had heart surgery. The departments were in different buildings but at least on the same campus.

The following day was Friday, not a good day on which to operate, what with locum doctors and agency nurses working the weekend. I needed the best team I could possibly muster, and as Julia was perfectly stable I decided we should do it on Monday morning. No fuss. Just another aortic valve replacement, but with a careful plan and the right back-up.

What makes a quick surgeon? Not haste or rapid hand movements. In fact, quite the opposite – being well organised, not doing unnecessary things, getting every stitch where it needs to be and not having to repeat anything. So quick surgeons don't move fast, it's just a matter of connection between brain and fingertips. You have to be born that way. No amount of training helps with this.

Now I needed to meet her. Oliver took me to her room on the cardiology ward. She was alone. No family there in the morning. As promised she was feisty and probing, but anxious about what I might say. Nervous that others had repeatedly argued for termination.

Her first words were, 'I want to keep the baby.' I replied that I wanted to keep the baby too. Then we had a working relationship.

So when would the operation be? I told her it would be on Monday morning, then described the type of valve we'd be inserting and the fact that she wouldn't need anti-coagulation. Clearly this was important for the later stages of pregnancy and delivery. I said that the valve would wear out and that she'd need another one in fifteen years, maybe less. But Julia wasn't thinking that far ahead. She just wanted this horrific intrusion removed from her ordered life.

'Can I go home for the weekend?' she asked, eager to make some arrangements and let work know.

'OK, but no riding and no exertion – of any kind! But you've got to stay right now until we've cross-matched your blood and the anaesthetist has been to see you.'

Oliver agreed that it was a good idea for her to go home and that there was no point insisting otherwise. Monday's anaesthetist was Elaine – I called her to explain the delicate situation and she came directly. As Elaine talked to Julia I went to warn the perfusionists and give them some literature, telling them what I wanted and emphasising that two lives were at risk this time.

When I next saw Julia at 7 am on Monday she was

perfectly calm. She asked me to save the deformed valve as it belonged to her and she wanted to keep it. Her whole family were in the room – husband, sister and her elderly parents, all there to lend moral support. I said I'd come back and talk to them later.

The arterial and venous monitoring cannulas were put in using local anaesthetic. I really didn't want to monitor the foetal heart rate. I'd done that before and found it a source of anxiety and distraction if the rate slows, given there's nothing we can do to change things when we're already taking the appropriate precautions. Elaine was careful to keep the blood pressure and oxygen levels up during anaesthetic induction. We checked the foetal heart rate before taking Julia through into theatre. It was normal – 140 beats per minute, twice that of the mother. The echo probe was pushed down the gullet and into Julia's stomach, ready to view the heart. We kept her covered with blankets until the last minute, to avoid cooling. Then off with everything. The little abdominal bump was there to remind the team to stay focused.

Soon she was painted with antiseptic and covered in blue drapes, with just the long, narrow furrow between her breasts left on display under the iodine-impregnated adhesive drape. We attached the electrocautery, defibrillator and pipes for the heart–lung machine to the operating table, and then we were ready to go.

My scalpel sliced through the skin layer – there was more bleeding than normal because of the hyperdynamic circulation – then the electrocautery cut through the thin layer of fat down to the bone. Now the saw up the middle of the breastbone – buzzz! The bit that turns the students' stomachs and makes them faint. It oozed marrow. More cautery through the remnant of the thymus gland, then into the pericardium. Elaine gave the heparin in readiness for cardiopulmonary bypass.

We inserted the cannulas through snares in the aorta and right atrium, then it was on to the machine. We stopped ventilating the lungs and the circuit took over, and instead of cooling we used the heat exchanger to keep her warm and high pump flow to keep both uterus and placenta happy. With a clamp across the aorta, in goes the cardioplegia fluid until the heart stops dead. Not really dead, but flaccid and cold, protected by having its metabolism halted.

I used a scalpel to slit open the aorta and expose the offending valve. It was not recognisable as a valve. Just as the echo indicated, it was a rock-solid volcano with a narrow slit. I cored it out in one piece with a different, sharp-pointed blade, then placed it gently in a bottle of preserving fluid, my gift to Julia. Next I sewed in the new biological valve with twelve separate stitches. This had been carefully constructed from the pericardium of a cow and was suspended from a plastic frame with a sewing ring. Now it was sewn on to where the old valve had been removed, a common and uncomplicated operation this time benefiting two patients – one present, one future. So far it was going well.

We sewed up the aorta and removed the clamp, warm blood flooding into her coronary arteries. The heart was reanimated by this life blood, first squirming in ventricular fibrillation followed by sudden spontaneous defibrillation. It lay perfectly still until I poked it, then contracted and ejected blood. After poking it again, normal heart rhythm began, the echo probe showing the artificial valve opening and closing. The way out of the left ventricle was now wide open for the first time in decades and thousands of small air bubbles dashed towards the needle. Routine and unexciting was precisely what we needed.

I told Elaine to start ventilating the lungs, check the blood gases and get ready to come off cardiopulmonary bypass. She rhythmically pumped air into the windpipe and the collapsed lungs filled with air and expanded – from flaccid and empty to inflated, pink and proud. They surrounded the heart, same as always, day after day. We stop life and start it again, making things better, taking calculated risks.

A pulse wave returned to the arterial pressure trace, now regular and strong. But I wasn't looking at the screen. I was watching the heart itself, still pushing out the last residual air bubbles. They floated upwards, straight into the right coronary artery, now obstructed by coalescent air. The right ventricle lost its blood supply and temporarily distended. No big deal. We increased pump flow and blood pressure to push the air through. The right ventricle contracted again, and all was well.

Now I wanted to get off the machine as soon as possible. I told the perfusionist to slowly ease off bypass to let Julia's re-plumbed heart take over. We'd been on the machine for just forty-nine minutes, and during that time we'd maintained high flow rates and normal temperature, doing our best to look after the uterus and its precious cargo. I heard 'Off bypass'. We remove the cannulas and reversed the heparin anticoagulant with protamine.

The cut surfaces were still bleeding, more than usual. My ADHD and irritable bladder kicked in and I felt it was best to let Mohammed finish off – cauterise the bleeding sites, put the drains and pacing wire in, make sure she's safe. We were trying to avoid blood transfusion with its negative effects, but at the same time we didn't want oxygen delivery to be compromised by too few red cells. Eventually we needed to give her two units of donor blood, and fresh frozen plasma with its clotting factors, followed by platelets, the sticky cells that plug small holes. Within an hour she was ready to transfer to intensive care, her bleeding under control.

Elaine and Mohammed escorted Julia out of the operating theatre complex, elated that everything had gone according to plan. Yet after all the preparation they were greeted by an inexperienced nurse. The intensive care unit had been warned like everybody else – and it was hardly the poor nurse's fault – but Elaine was irritated. What was the plan to look after the baby? When was the most likely time for the baby to die? What would we do if Julia had a torrential bleed down below? Blank faces, wide-eyed nurses, stupefied junior doctors. So get an experienced team together and get focused. I was unaware of all this at the time but Elaine was right. Experience matters a great deal in high-risk situations, and in this case two lives were at risk.

Julia's blood pressure was on the low side. Her blood vessels were more dilated than usual because we'd kept her unusually warm on the bypass machine, but we couldn't give her the standard drugs we used to increase pressure as they'd constrict the blood vessels to the uterus and placenta. Nor could we allow an average blood pressure less than 70 mm Hg. The solution was in the guidelines. Everybody had it, but did anyone in the intensive care unit read it? No point saying anything, otherwise there would be a complaint.

When I returned I asked Mohammed to stay with her. Oliver used an ultrasound machine to image the foetal heart, which was still beating away at around 140 beats per minute. So far we had a viable foetus and no uterine contractions, so I told them to let her wake up, take her off the breathing machine and get rid of the sedative drugs. Then her blood pressure would rise spontaneously. I left to operate on my next patient with a parting shot – 'You're looking after two people now, not just the one you can see.'

Julia woke up quickly and eventually had the tube removed from her windpipe. She described being awake with the tube in as the worse part of the whole experience. I met Oliver at seven o'clock the following morning to image the foetal heart, still bounding away at 140 beats per minute. Not just that, but the foetus was doing somersaults in the womb. And Julia's heart was working well with the new valve – she had warm feet and a good volume of urine in the catheter bag. Medicine is the only profession where piss is a cause for celebration. Nonetheless I remained uneasy, as her blood pressure was on the low side. Our incomplete understanding of heart surgery in pregnancy meant that we didn't know if it mattered at this stage, but we still didn't want to use drugs that would compromise the placental blood supply.

On awakening, Julia's first question was whether her baby was still OK. We reassured her that it seemed to be, but we looked forward to finding a strong heartbeat in another twenty-four hours. By then I felt we'd be in the clear. Later that morning we took the drain out of her chest. She was desperate to get back to a single room on the ward but I wanted her blood pressure and oxygen levels monitored for another twenty-four hours. We moved her to a quiet isolation room normally used for septic patients.

The next day the foetus was still the same, moving and with a normal heart rate, but Julia was uncomfortable. The second post-operative day is always the worst, the first day bringing the euphoria of survival, the next just pain. Unfortunately, for the baby's sake, we couldn't give her a heavy analgesic regime.

We operated on Monday. By Friday Julia was bored, comfortable and insisting on going home. We couldn't stop her. A concerned Oliver called her each day for the next week, then saw her regularly in the outpatients department. Foetal ultrasound scans showed normal growth and activity. Five months later, in January 2016, she delivered a healthy baby boy weighing nine pounds – her miracle child, once destined to be a heap of scrapings in a stainless steel dish. We changed all that, Oliver and me. Welcome to the world, Samson. Strong man!

16

your life in their hands

Have a heart that never hardens,
And a temper that never tires,
And a touch that never hurts.

Charles Dickens, *Our Mutual Friend*

IT WAS 2004, almost fifty years since that television programme planted the seed in my cerebral cortex – *Your Life in Their Hands* from the Hammersmith Hospital, the programme that shaped my destiny. The BBC had called my office and spoken with my secretary Dee. She was excited when I popped by in between cases. Would I consider doing a show for them, a full hour of prime-time television? They were looking for a brain surgeon, a transplant surgeon and a heart surgeon. The name of the series, *Your Life in Their Hands*.

The distinguished producer and his female assistant came to Oxford to talk through the implications, explaining that the filming could be intrusive for a while. They'd spend six months with me at the hospital and at home, meeting the patients and interacting with my family, so viewers could appreciate how it felt to be a heart surgeon. Life at the sharp end, in my case very sharp.

They wanted me to implant a Jarvik 2000 for the cameras and asked whether I could find them an appropriate heart failure patient to follow, before, during and after the operation. Of course, they would feature other cases. They'd like a baby case and other dramatic, high-risk stuff – cutting-edge, inspirational surgery in real time, whether the patient lived or died. They'd do the filming and decide what material to use. No pressure, then.

From their background research they knew that I regularly operated live for surgical audiences, and that I was flamboyant and a confident performer, not easily intimidated. If I agreed to do it they'd make the arrangements with the hospital. At the time we had a chief executive who actually talked to us, a likeable guy who periodically emerged from his ivory tower to visit the worker bees – I was in no doubt that he'd agree to it. Now I just had to let my family know that a film crew would be coming home with me after work. And meeting me in the mornings. And interviewing them. How was it to live with a heart surgeon? Good question!

It soon became normal to have a film crew on my shoulder. Many operations were recorded: premature babies with holes in the heart, young adults needing massive surgery for Marfans syndrome and a middle-aged lady needing her fifth aortic valve replacement, a difficult procedure that eventually took twenty-four hours. Things went dramatically wrong on camera but she survived. Of course they used that material.

They filmed me jogging with Mark and watching Gemma play golf for Cambridge University. But after several months there was still no suitable candidate for the Jarvik 2000. Eventually I called Philip Poole-Wilson at the Royal Brompton Hospital. It took him less than a week to locate the ideal patient, a delightful fifty-eight-year-old Scotsman who'd already been turned down for transplantation in Glasgow. Jim Braid was very much in the mould of Peter Houghton. He was dying but desperately wanted to survive long enough to see his daughter graduate and get married. As the clock ticked on relentlessly it was clear that he wouldn't make it.

A long time had passed since Jim's transplant assessment and we needed up-to-date information. Philip brought him down from Scotland and admitted him to the Brompton. He needed repeat right and left heart catheterisation, detailed echocardiography and lots of blood tests. I was mindful of the fact that we were still paying for all of this out of charitable funds. The NHS wouldn't pay – they'd written him off like Peter and the others. I was his only chance.

Glasgow were correct about him being unsuitable for a heart transplant. The blood pressure in his lungs was too high, although Jim's own right ventricle was used to it. It was his left ventricle that was going down. He had the same problem as Peter – dilated cardiomyopathy. Nor were his kidneys working well enough to tolerate the immunosuppressive drugs that a heart transplant would need. A left ventricular assist device would take over from his flabby, failing heart. Not only that – it might help to rejuvenate it. Possibly. The echo showed it certainly couldn't afford to get any worse. It was now or never. We couldn't risk letting him go home to Scotland.

I took the excited BBC team down to the Fulham Road to meet Jim and his wife Mary. Peter Houghton came down from Birmingham and was on great form, still raising money so that others could have pumps. It had now been almost four years since his implant and he was approaching the world record for survival with any type of artificial heart. He was pleased to counsel Jim and Mary, doing it professionally and keen to be regarded as part of the team.

They were justifiably nervous but eager to proceed, suitably impressed with the technology. Moreover, Jim was a great character, perfect for television. He shuffled down the corridors, head bowed, panting for breath, his nose and lips blue, and although he could barely speak he'd still joke for the camera. 'It's great to be down in London with these Ferrari mechanics, not like the Ford Escort boys up north,' he'd say. That resonated with me.

It was good to be back at the Brompton. As most of the original intensive care team in Oxford had now moved on, I asked Philip if we could do the implant in London, which pleased him enormously. First I needed to engage with the senior surgeon, Professor John Pepper. He was happy to help out, so we planned the implant for the following week. Rob Jarvik agreed to fly the pump across from New York at short notice and my Oxford colleague

Andrew Freeland would come to help with the skull pedestal.

Now we had the patient, the pump and a top team – the producer's dream. All we needed was a successful implant with the cameras rolling, and Jim must survive. But as the Brompton anaesthetists emphasised, he wasn't fit for an anaesthetic. Nevertheless, the hospital remained keen and supportive, and we didn't have to fight the management to do it. They'd never implanted a left ventricular assist device before and would have been disappointed if we hadn't gone ahead.

Five thirty in the morning, dark and cold. The film crew picked me up by taxi and we headed into Oxford to find Andrew. He was wandering down the Woodstock Road carrying instruments to screw the plug into Jim's skull. We headed down the M40 doing an interview in the car.

'How do you feel about operating in a different hospital?'

'Excited. I've operated everywhere from Tehran to Toronto. An operating theatre is an operating theatre and I've a good team. As Baldrick would say in *Blackadder*, "We have a cunning plan!"'

'And how do you feel about the fact that he could easily die? Are you nervous?'

'Absolutely not. Jim will be dead in days if we don't try. No one else is going to help.'

'Do you think that the NHS should pay for these pumps?'

I answered that with a question of my own. 'Should a First World health-care system use modern technology to prolong life? Or should it let young heart failure patients die miserably, like in the Third World?'

The BBC liked that answer, but they didn't broadcast it in the programme. Too controversial and intimidating.

We reached the Brompton at 7 am, and I took Andrew and the crew directly to its deserted canteen. Little had changed since my day. They still did a good breakfast, so I helped myself to the healthy option: sausage, bacon, black pudding, fried egg and fried bread. Andrew followed suit. As we sat together the cameras started rolling. It was what the producer was waiting for. Heart doctor eats huge pile of fried food – wall-to-wall cholesterol.

Me: 'This is great. I never get this at home.'

Andrew: 'What would your wife say about that?'

Me: 'Don't care!'

This encounter turned out to be what everyone remembered about the programme. When my brain surgeon friend Henry Marsh did his episode they filmed him cycling to work through the streets of London – without a crash helmet! When asked he simply stated, 'Never wear one. It wouldn't save me!' The BBC wanted characters and they got characters.

John Pepper came down to meet us. Under the circumstances we were a very relaxed group, perhaps not what you might expect but good for Jim. Stressed surgeons do not function well, as numerous studies have shown. Stress impairs judgement and makes the hands shake. In fact stress is killing my profession.

We went to the ward to see Jim and Mary before they brought him down. Jim was excited, Mary petrified. Was this her last goodbye, the end of their journey together? And would it be the high road or the low road back to Scotland? I did what I always do at such moments – reassured them that everything would be fine. Not that I knew that. I just wanted them to go into the operation with confidence. With the cameras rolling we were all in this together.

There was an air of busy excitement in the operating theatre – nurses setting up trays of glistening instruments, perfusionists assembling their heart–lung machine, technicians jealously guarding the artificial heart, only to be revealed at the crucial time. But no Lord Brock's boots this time. I was my own man.

Now uncovered, poor Jim was obviously emaciated with the heart failure. The left side of his head was shaved, ready for the skull pedestal and power line. He was about to become battery-powered. Beginning with needle and guidewire then small stab wounds, John inserted the pipes for the cardiopulmonary bypass machine into the main artery and vein of Jim's left leg. This was more sophisticated equipment than mine. I was learning something.

Once Jim's chest had been prepared with iodine and draped with adhesive film, Andrew exposed the surface of his skull while I sliced open the ribs, the camera panning from one site to the other. About a litre of straw-coloured fluid poured out of Jim's chest – heart failure juice. I could then see the hugely dilated left ventricle through the pericardial sac.

I started to tunnel the pump's electric driveline out through the apex of the chest and into the neck, avoiding the perilous blood vessels and nerves to the left arm.

Having penetrated through the neck I delivered the miniature plug on the end to Andrew. He passed this through
the middle of the titanium skull pedestal, then screwed the
titanium onto the skull behind the ear. Rigid fixation, it
was called, so that the external power line could be
plugged in securely. It all looked fascinating on television,
but we still hadn't reached the tricky bit.

As I opened the pericardium clear fluid spilled out. The
pale, distended left ventricle just twitched – you couldn't
dignify it with the word 'contraction'. I urged the cameraman to focus in on it as I was about to sew on the pump's
restraining cuff. Each time the needle pierced the muscle
the heart twitched and threatened to fibrillate. Irritating,
because I was trying to do the implant without starting
the heart–lung machine. This would reduce the risk of
bleeding at the end of the operation. But Jim was too
unstable. Before the cuff was secure the heart did fibrillate. No blood pressure but no problem. We just started
the bypass machine and emptied the heart.

Now came the exciting sequence for the film – carving
the hole in the apex of the heart to insert the Jarvik 2000.
First I made the cruciate scalpel incision, during which
blood always spurts out. Next we cored out a circle of
muscle with a cork bore device, causing blood to pour
into the pericardium. This was halted when we inserted
the titanium pump into the heart, and with a professor of
surgery assisting me it all went smoothly. Andrew
connected the external power cable to the skull pedestal
and we switched Jim on, slowly at first until blood
expelled the air from the Dacron graft.

As usual air came fizzing and frothing out of the needle, red bubbles forming on the white tube. Visually it was extremely satisfying. I instructed the perfusionist to cut back on flow so we could fill the heart before turning up the Jarvik 2000 impeller speed. The last few air bubbles spluttered out from the highest part of the ventricle, apex uppermost. It was simple physics and done without thinking. There was lots of chemistry going on at the same time – optimising the potassium level and neutralising the lactic acid with sodium bicarbonate – as well as biology, with the electrical defibrillation of the quivering muscle to provide a stable heart rhythm. My three school examination subjects were not wasted.

For many viewers it was the engineering that proved the most exciting part: an electric plug in the head and a turbine in the heart, spinning at 12,000 rpm without damaging the blood cells, and no pulse in the circulation. I maintained a continuous commentary for the television while throwing out instructions to the anaesthetist and perfusionists. 'Start ventilating the lungs. Reduce your flow. Turn up the Jarvik.' Detailed coordination by a guy who wouldn't lift a car bonnet and couldn't use a computer. No one quite believed how well it had all gone.

Were we pleased for Jim or focused on making great television? The honest answer is both. Naïvely, I thought that if the public could see his miraculous recovery there would be pressure to treat patients with these techniques on the NHS. We couldn't sustain a charitably funded programme any longer – it was second-hand-shop health care. Poole-Wilson had this in mind too.

We wanted to do a proper clinical trial by randomly allocating dying heart failure patients to a ventricular assist device or continuing medical treatment. We knew what the outcome would be: symptom-free extended life versus inexorable deterioration and death. We didn't consider that fair to those who didn't get a pump, but the devices would never be approved for NHS use without a trial. Only the British Heart Foundation had enough money to support this endeavour and they turned us down. At the time it couldn't be done in the States, either. They were waiting to see the long-term outcome of pulse-less patients before agreeing to it, so all eyes were on us.

Jim separated easily from the bypass machine. This was the most taxing part for the Brompton anaesthetists. It was the first time they'd managed a patient with continuous blood flow. An average flatline blood pressure of 80 mm Hg was optimum, although for any other heart patient this would be regarded as unreasonably low. Vasoconstrictor drugs would normally be used to raise it beyond 100 mg Hg, but Jim needed a counterintuitive approach.

We gave him vasodilator drugs to reduce blood presure. The lower the vascular resistance, the greater volume of blood the Jarvik 2000 would pump. His organs needed sufficient perfusion pressure, but 70 mm Hg to 90 mm Hg was just fine. The kidneys, liver and brain work normally at this level, tiny capillaries supplying the tissues – there's no pulse in capillaries even when the arteries are pulsatile. We learned all this through trial and error. It worked in the laboratory, so it should be fine on the wards, albeit a source of fascination for the Brompton team and the film crew.

Andrew closed the scalp and neck incisions, then took off for Oxford. He had a busy clinic that afternoon – snotty noses and wax-filled ears, not artificial hearts. John removed the pipes from Jim's groin, and I inserted the chest drains and started to close the chest wound, meticulously cauterising all the bleeding points. His scalp was still oozing, so I put a couple of extra stitches to the skin, then cleaned the blood off the skull pedestal. Today cosmetics mattered. We needed clean white dressings and empty drains, every single spot of blood cleaned away.

Nostalgically, I remembered my first heart operation in this very same operating theatre. I'd been wearing Lord Brock's boots, and I recalled that when I pushed the saw through the poor lady's sternum and into her heart Matthias Paneth strolled through the theatre doors in his pinstriped suit and exclaimed, 'Westaby, what have you done this time?' Now it was me in charge.

The cameras kept rolling as Jim was wheeled off to intensive care. I glanced back into that operating theatre for the last time. There were pools of blood under the table, glistening bright red under the lights, and a puddle of urine where the catheter bag had leaked. The perfusionists were folding their redundant tubes into a yellow plastic container, bloodied green drapes were stuffed into clear plastic bags and the nurses in their theatre blues were disposing of the redundant white swabs – all the colours of the rainbow, an artist's dream.

It had been a historic day. The backstreet kid from

Scunthorpe had implanted an artificial heart at the Brompton for the television programme that took him there in the first place, fifty years before.

Once Jim was safely connected to the ventilator and monitors we went to find Mary and her daughter, cameras in tow. There was no escape. Drama they were after and drama they were determined to find. We took the family to see him. The surroundings in critical care are always intimidating, this time particularly so. Jim's scalp was shaved and a black power cable dangled from his head, life dependent on a battery.

We explained everything to them but they already knew most of it from Peter Houghton, who was now on his way to the hospital. But they couldn't see the electric plug under Peter's hair. It was rather more alarming when they confronted it head on. I handed Jim's daughter a stethoscope and placed the listening end over his heart. A look of surprise lit her face. She could hear the continuous whine of the spinning impeller that would keep her dad alive. I pointed to the cardiac output monitor. The implanted device was pumping four litres of blood per minute and consuming seven watts of power via the controller and battery. I could turn Jim's blood flow up or down. It was easy – just one single knob. The producer loved it. This was much more exciting than brain surgery. Drilling small holes in the head and sucking out bits of tumour? That needs a different personality type altogether.

Jim stayed incredibly, boringly stable. He didn't bleed, whereas Peter and others had lost gallons. John, Philip and

I wistfully discussed other potential patients. Where could we get the money? I could raise enough for a few more pumps but not a full-scale trial. The discussions ended up where they needed to, in the pub, cameras and all.

When I wandered back to intensive care Peter Houghton was with the family, beaming like the Cheshire Cat. It was important for him to have what he called 'cyborg companions' – battery-driven people making a new life for themselves, Dr Frankenstein's monsters with a metal bolt sticking out their skull. For me this was a happy scene and I felt that one day all life would be this way. On that curious note of fantasy, I decided to go home to Woodstock. The longer I stayed at the Brompton, the more I wished I still worked there. It was a 'can do' environment – a famous old hospital wanting to do new things, not seeking reasons not to do them.

I operated in Oxford the following day, then headed back to London. Jim had been taken off the ventilator, the tube was out of his windpipe and he was chatting to Mary, back in the land of the living. He looked completely different, animated and radiating joy, his nose and ears pink not blue. The pump was pushing out five litres of blood per minute with absolutely no pulse on the arterial pressure trace. And there was a litre of urine in the bag, this liquid gold, meaning that his kidneys were happy.

By now the camera team were in the pub. I asked the intensive care doctor whether he'd prescribed warfarin yet. It was all done and there was nothing for me to add. This dead-end heart failure patient was recovering rapidly, with no immunosuppression or any of the other poisons that a heart transplant patient needs. What's more, Jim's own right ventricle was coping well with the extra blood flow. So when I returned to Woodstock it was with a sense of deep satisfaction.

I saw Jim several times before he returned to Scotland. Philip greatly reduced his heart failure drugs, in particular the water tablets that make every patient's life difficult, and the family had no problem getting used to the pump, changing the batteries regularly and plugging it into the mains overnight. Jim's ankles slimmed down, he was no longer breathless and he could lie flat for the first time in months.

Weeks later he was there when his daughter graduated, with a glass of champagne in his hand. Then the BBC filmed him walking along a Scottish beach with Mary at sunset – a happy man, breathing easily, reflecting on his journey – and this poignant scene was used to close the programme. The *Your Life in Their Hands* series won a prestigious award for Best Television Documentary, and I was proud to have played my part in that. It was a high point in my career.

Only rarely did Jim return to the Brompton for a check-up. The local hospital and his GP became familiar with the technology and were happy to look after him. But then came dismal news from Scotland, shortly before Christmas. Jim had gone to visit a friend without taking a spare battery with him. He was enjoying life and his mind was on other things. The 'low power' alarm went off on the controller, meaning that he had twenty minutes to change the battery before the power went off altogether.

Jim didn't make it home. His own heart had not recovered sufficiently to see him through. When the battery expired Jim died too, his lungs filled with fluid. It was desperately sad after three years of good-quality extra life. For me this catastrophe illustrated just how effective these devices can be. It was a tragic loss.

Time passes. Before I knew it, it was 2016. By now I'd had a lifetime in cardiac surgery. How much longer did I want to spend doing this? The trouble was that I remained good at it, a compulsive operator who would take on the difficult stuff and, after thirty-five years, vastly experienced in a way new surgeons could never be. Did I owe it to the patients to stay? Or to my family to quit, to move to an easier job?

My personality and retirement were not in the least bit compatible, but my right hand had become deformed. The fascia in my palm – where the scrub nurse slapped the metal instruments – was contracting and I was developing a claw hand, known as Dupuytren's contracture. Now I couldn't even greet people properly, as my hand was warped into the position in which I held the scissors, the needle holder, the sternal saw. It was a true occupational adaptation, and one that would ultimately force a decision. Then, as for many ageing surgeons, bending over an operating table for hours on end eventually took its toll on my spine. As I used to instruct my registrars, 'Please take over – my back's bad and the front's not so good either.' Yet no physical ailment was as debilitating as hospital bureaucracy, not being able to operate, no beds, not enough nurses, junior doctors on strike. In addition, there was the 'statutory and mandatory' training, where I had to sit in a classroom while a paramedic taught me how to resuscitate, or taking a quiz on how to prescribe insulin or cancer drugs – things that I never ever have to do – or writing my personal development plan at the age of sixty-eight. It was all time wasted, when I really should have been up to my elbows in someone's chest, doing some good.

The fire alarm went off in the operating theatres recently, right in the middle of a valve operation while the patient was still on the bypass machine, their heart cold and flaccid, the prosthetic valve half sewn in. An administrator poked her head round the door and said, 'The fire alarm has just gone off. We don't think there's a fire but we have to evacuate.'

I just said, 'Right then, I'm off.' The look on her face was priceless. I followed up with, 'You rush off then. Save yourself. But please leave us a bucket. We'll piss in it and put the fire out!' One can only tolerate so much. My whole profession had lost direction.

afterword

Don't cry because it's over.
Smile because it happened.

Theodor Seuss Geisel (Dr. Seuss)

AFTER I QUALIFIED in medicine in 1972, the old Charing Cross Hospital closed down and relocated. When the last patient left the famous landmark on the Strand, many of us students walked around the empty shell to reminisce about our training. I went back to that rickety old lift and up into the eaves, and for one last time I opened the green door into the Ether Dome. The electric lights still worked but all the dusty antiquated equipment had gone. I walked tentatively across the boards to gaze down into the operating theatre, just as I'd done six years before. Sure enough, that last spot of Beth's blood was there on top of the operating light – black, ingrained and inaccessible. They never did succeed in washing her away.

Beth continued to come to me in the dead of night, particularly during the bad times, of which there were many. The baby was now in her arms, behind it the brutal metal retractor embedded in her frail chest, her dead heart empty and still. She'd walk towards me, ashen white, with her piercing eyes wide open and staring at me, exactly as they were on that day. Beth wanted me to be a cardiac surgeon and I didn't disappoint her. I was good at it. Yet despite my best efforts some patients took the fast track to Heaven. How many, I really don't know. Like a bomber pilot I didn't dwell on death. It was more than three hundred, fewer than four hundred, I guess. But Beth was my only ghost.

June 2016. It was an astonishing fifty years after I tentatively passed through the doors of the dissecting room as a nervous young student to start cutting on a wizened, greasy and embalmed human body. Now I was standing on the podium in the Royal College of Surgeons, holding court at a meeting for heart surgeons in training. The organisers were parading me as a role model – a pioneering heart surgeon who had survived the course without being sued or suspended. An increasingly rare species. My talk was about the illustrious history of the heart–lung machine and circulatory support technology, celebrating the great men and daring deeds that I grew up with, not to mention those I did myself.

As the next lecture began I attempted to slip out unnoticed. But there was a flurry of activity behind, a scramble of eager young men who wanted their picture taken with me. I found that flattering. We posed in front of the marble statue of John Hunter – legendary surgeon, anatomist and bodysnatcher – in the entrance hall. I always felt uneasy at that particular spot. It was where I learned that I'd failed my exams – on more than one occasion – when my name wasn't read out. When many of us walked away in shame.

Even my eventual triumph was painful. It was the time I took the oral examinations with a badly fractured jaw that kept me quiet. On a grim winter's afternoon in Cambridge I'd been sitting in the Addenbrooke's accident department covered in mud following a misjudged rugby tackle. Still in my rugby kit, I was waiting to see the orthodontic surgeon when an ambulance brought in a young motorcycle-accident victim who was bleeding to death into his left chest. There was no time to call the cardiac surgeons from Papworth Hospital, so the casualty officer and nursing sister, who both knew I'd worked there, asked me to intervene before it was too late. I opened him up wearing grubby shorts with muddy knees, spitting out my own blood into the scrub sink.

This bizarre story went viral and there were Cambridge surgeons at the examinations. Maybe it even helped. Yet ultimate success hadn't dimmed those memories. I detested the cardboard elitism, with the examiners dressed in bright red gowns sweeping around the pillars – 'Flash Gordon outfits', I used to call them. Now the Royal College had become an institution that tacitly supported the 'name and shame' culture, that endorsed the public release of named surgeons' death rates, that tipped its hat to the politicians who ruled health care in preference to defending its members.

How things had changed since my day. Despite the hardships, when we made it into heart surgery we felt ten feet tall – proud, bullish, like fighting cocks. The world was our oyster, we'd reached the top and people respected us. By contrast, these trainees were downtrodden, defensive, uncertain of themselves. The mood in the College was sombre.

One earnest young man from the Middle East wanted to talk. His hospital was under investigation for borderline results, his mentors – whom he respected – were being chastised in the newspapers and as a result he wondered whether he should continue. Was it worth the struggle, or should he give up and go home to his family? I told him that I'd once operated on a sick blue baby in Iran, the child of a politician back in the bad old days after the Revolution. At the time, although I was really concerned for my own safety had the child not survived, I stuck my neck out because the patient had no other option. So my first piece of advice to him was, 'We're here for the patients, not for ourselves. We may suffer for that but we'll rarely regret it.'

We left the gloom of that historic building and walked down to the Strand in the sun. I asked him why he'd chosen heart surgery in the first place and he told me it was because his sister had died from a congenital heart defect. He wanted to operate on children but this already seemed a 'bridge too far'.

As we passed the Savoy I explained my own background, about losing my grandfather to heart failure and wanting to find a solution. If a backstreet kid from Scunthorpe could do it, so could he. Then I told him about Winston Churchill, whom I often talked with in the graveyard at Bladon. How in the dark days of the Second World War and during his 'black dog' hours he never gave up, and how I didn't give up after the débâcle of my own first heart operation. So my second piece of advice was, 'Follow your star. Do it for your sister.'

We turned off from the Strand and went past Rules restaurant in Covent Garden. As an impecunious student I'd seek to impress potential girlfriends there, then starve for the rest of the month. I told him not to be afraid to take risks. Sometimes they pay off handsomely. We walked for another couple of hundred yards and there was the entrance portal of the old Charing Cross Hospital, my glorious medical school now turned into a police station. I described to him the Ether Dome and the operation that haunted me, a catastrophe that could have changed my life. But it didn't. It made me more determined to press on against the odds. So perhaps one last thought for him: 'The past is the past. Put it behind you. It's tomorrow that matters.'

The young man was grateful. Taking the time to talk made a difference to him. Perhaps he felt as I did in America when Dr Kirklin told me to take the difficult route and operate on children, or when Dr Cooley first showed me an artificial heart. As he turned to go back to the conference he went to shake my hand. From his quizzical look I could tell he was surprised that the hand was badly deformed. Until recently it hadn't interfered with my operating. I'd been advised to have surgery long before but typically had ignored it, concerned that it would end my surgical career. Now it had gone too far. I could no longer grasp the instruments without dropping some of them and I couldn't shake hands without people thinking I was a member of some secret society.

At that point I conceded that my operating days were over. I'd never get back to complex surgery. Instead my focus would be on our new stem cell research and the ventricular assist device we were developing – plenty to engage with but different, research with the potential to change millions of lives. Just a few weeks later I quietly disappeared from the hospital and had my right hand operated on. Normally my plastic surgery colleagues would have done it using a regional nerve block with me awake, but they didn't want the interference. Frankly, I was pleased to be asleep as I really didn't enjoy being on the other side of the fence. And for me it wasn't just an operation. It was the end of an era.

acknowledgements

MY MENTOR IN THE UNITED STATES was the great Dr John Kirklin, who launched open heart surgery with the heart–lung machine. Towards the end of his distinguished career he wrote:

> After many years of cardiac surgery, with many tests and challenges, and after many deaths that could not then be prevented, we tend gradually to become a little weary and, in some sense, infinitely sad because of life's inevitabilities.

I wrote this book because I've reached that same point in a career spanning the rise and fall of the NHS. Hence my acknowledgements are as emotionally charged as the rest of the text.

Heart surgery has been in turn a difficult road and a lonely destination. In the 1970s and 80s we really did work constantly. In the States it was 5 am ward rounds, call the boss at 6 am, operate all day, go to the laboratory in the evenings, capped by night-time vigils beside the

intensive care bed. It was not so different at the Brompton
or Hammersmith hospitals in London.

The pioneering early days were highly competitive, and
junior heart surgeons were the thrusting young blades of
the medical world. I was lucky. I succeeded because early
in my training I learned from the great men: Roy Calne,
John Kirklin, Denton Cooley, Donald Ross, Bud Frazier
and many more. I understood what was necessary to
move the specialty forward. For me it took relentless
effort and lateral thinking, then sheer guts to go with the
blood.

This ruined any aspiration towards a normal family
life. We were not normal people. Most rational young
men would be paralysed by fear at the thought of carving
open someone's chest, then stopping, opening and repairing their heart. But I did this day after day. We were
fuelled with testosterone, driven by adrenaline. Few of us
stayed married in our youth, and many of us subsequently
harboured deep regrets.

I was always sorry for the distress caused to my first
wife Jane and eternally grateful for my talented daughter
Gemma, now a Cambridge-educated human resources
lawyer. While I spent many hours striving to save other
people's children, I never spent enough time with my own.
This book goes some way to explaining why I was so
preoccupied. It also gives me the opportunity to empha-
sise that nothing ever mattered more to me than them –
and the rest of my precious family. My only sibling David
attended the same grammar school in Scunthorpe but did
go to Cambridge. He read medicine at Christ's College,

then followed me to Charing Cross and became an eminent gastroenterologist in London.

Inevitably, I met my soulmate over an open chest in the accident department, blood everywhere, drowning in sheer desperation. Sarah was the kindest accident and emergency sister I ever met. The daughter of a Battle of Britain Spitfire pilot, she was never flustered, nothing was too much effort. The lad died, and when I couldn't face telling the family she dealt with it. She did the same for others, time and time again. A free spirit from Africa, she made no distinction between tramps and politicians – they were all valued people to be treated with respect. I ruined her relationship and she suffered considerably for it. But she went on to give me unquestioning and unfailing love and support for the past thirty-five years, particularly through those difficult times. Mark came along ten years after Gemma, a sportsman and adventurer who took himself off to South Africa to train as a game ranger.

It was a struggle to build the Oxford programme. The hard work was done by a handful of dedicated personnel who took the cardiac centre from fewer than one hundred operations a year in 1986 to more than 1,600 in 2000. Our productivity was allied to innovation, and the team was replete with skilled surgeons and cardiologists, supportive anaesthetists and perfusionists, and superb nurses – too many to name, but I'm grateful to all of them.

We could never have started the paediatric and artificial heart programmes without the support of one visionary hospital chief executive, Nigel Crisp, who went on to run the whole NHS and now deservedly sits in the House of

Lords. Much of the artificial heart work was undertaken with charitable funding. In that context certain individuals and organisations were very generous. These include Heart Research UK, Sir Kirby Laing, Jim Marshall of Marshall Amplifiers (to whom I was introduced by my patient, the entertainer Frankie Vaughan), Christos Lazari, and the TI Group courtesy of Sir Christopher Lewinton and David Lillycrop. I would also like to pay tribute to Professor Philip Poole-Wilson, past president at the European Society of Cardiology, who helped us greatly with the Jarvik 2000 Heart Programme. Philip sadly passed away suddenly on his way to work at the Royal Brompton Hospital.

Eventually, when I was the only remaining paediatric surgeon, we lost children's heart surgery. Then I had to move the artificial heart research away from Oxford.

I'm grateful to my friend Professor Marc Clement, head of both the Institute of Life Sciences and the Business School at the University of Swansea for providing us with a laboratory and an engineering team. We met serendipitously through my famous artificial heart patient Peter Houghton, who, with Nicki King, worked tirelessly to raise charitable research funding. Under the corporate banner 'Calon CardioTechnology' we now have an implantable British ventricular assist device to compete with the American pumps, all of which cost the same as a Ferrari! Stuart McConchie, past chief executive of the HeartWare Company and Jarvik Heart, came to help us with that.

The Welsh connection put me in contact with the Nobel Prize-winner Professor Sir Martin Evans of Cardiff University, who first isolated foetal stem cells. With his colleague Ajan Reginald and the company Celixir he has worked on a heart-specific cell for regenerative medicine. With pumps and cells we aim to create a definitive alternative to heart transplantation.

Despite a degree in biochemistry and a PhD in the bioengineering of mechanical hearts I'm a computer-illiterate technophobe who's unable to perform the simplest repair on a car. So I've relied on good old-fashioned secretaries. For the past ten years Sue Francis has kept me afloat. We'd both be in the office before 6.30 am. Our Portakabin window looked directly onto the noisy twisted pipes of an air-conditioning plant, like an apocalyptic scene from Banksy's Dismaland. In summer, flying ants ate through the window frames, then in the winter the cold rain seeped through the holes. I spent long, restless nights there, scrunched up on a small sofa, afraid to go home in case my patient deteriorated. Besides my patients, world-famous people visited that office – Christiaan Barnard, Denton Cooley, Robert Jarvik, even David Cameron, our last prime minister. All were bewildered by the modesty of an NHS heart surgeon's headquarters. But between us Sue and I achieved great things; she took home and typed hundreds of publications, not to mention this book.

In that context I'd like to thank John Harrison, who published some of my surgical textbooks. John encouraged me to write for the public and introduced me to my agent Julian Alexander, who made this book happen. It was also a pleasure to work with the expertise of Jack Fogg, Emily Arbis, Mark Bolland and the team at HarperCollins. I would also like to thank my medical artist, colleague and friend, Dee McLean, for her superb illustrations.

So what happened to heart surgery in the UK? After multiple hospital scandals the NHS in England decided to publish individual surgeons' death rates. Now no one wants to be a heart surgeon. And who would, with the long, taxing operations, the anxious relatives, and the nights and weekends on-call? It's a system entrenched in nonsensical bureaucracy, with the reward of public exposure for a run of bad luck. Already 60 per cent of the UK's children's heart surgeons are overseas graduates.

Ultimately the stars of this book are my patients, but I fear that few of the dramatic cases would now reach an operating theatre in the UK. In the final analysis a profession that dwells upon death is unlikely to prosper, undertakers and the military apart. As Dr Kirklin emphasised, death in cardiac surgery is inevitable. When a surgeon remains focused on helping as many patients as his ability will allow, some will die. But we should no longer accept substandard facilities, teams or equipment. Otherwise patients will die needlessly. The comedian Hugh Dennis is not noted for his empathy. On the satirical BBC programme *Mock the Week* he offered an alternative ode to Dr Kirklin's thoughtful statement:

Roses are red, Violets are blue.
Sorry you're dead, What can I do?

The answer? Bury the blame-and-shame culture and give us the tools to do the job!

glossary

AB180 ventricular assist device: a temporary centrifugal blood pump that was originally implanted into the chest. Now known as the Tandem Heart, an external blood pump used in cardiogenic shock.

acute heart failure: the left ventricle fails rapidly and cannot sustain sufficient blood flow to the body. The lungs then fill with fluid. Usually caused by myocardial infarction or viral myocarditis and has a high mortality rate. *See also* shock.

angina: crushing pain in the chest, neck and left arm due to limitation of blood flow to heart muscle in coronary artery disease. Typically comes on during exercise. If it comes on at rest it may warn of a heart attack.

angiogram: cardiological investigation where a long catheter is passed through the blood vessels into the heart. This allows blood pressure to be measured in the cardiac chambers and dye to be injected to visualise the coronary arteries or aorta.

aorta: large, thick-walled artery that leaves the left ventricle then branches to supply the whole body. The first small branches are the coronary arteries, which supply blood to the heart itself.

aortic stenosis: narrowing of the aortic valve at the outlet of the left ventricle, restricting blood flow around the body. Can be caused by a congenital anomaly or degeneration in old age.

arteries: the blood vessels that convey blood to the organs and muscles of the body.

blood pressure: pressure within the large arteries. Normally measured by a cuff and stethoscope or a cannula inserted into an artery. Normal blood pressure is around 120/80 mm Hg. The higher figure is when the left ventricle contracts; the lower, when it relaxes.

bridge to recovery: the process whereby a ventricular assist device is used to sustain the circulation and rest an acutely failing heart pending recovery from a reversible condition. If the heart does not recover, a limited-duration pump can be replaced by a long-term implanted device.

bridge to transplant: the process whereby a ventricular assist device is used to prevent death from heart failure until a donor heart can be found. At the time of transplant the pump and diseased heart are both removed.

cannula: a plastic tube inserted into the heart or a blood vessel to carry blood or fluid.

cardiac catheterisation: a long fine-bore catheter is passed from the groin or wrist up into the heart or coronary arteries. Contrast medium is injected rapidly to demonstrate the internal anatomy of the heart or blood vessels. The catheter is also used to measure pressure within the chambers.

cardiac tamponade: a condition that occurs when blood or fluid accumulates within the pericardial sac under pressure, preventing the heart from filling.

cardiomyopathy: heart muscle disease. There are several causes, which may be impossible to define, thus the term 'idiopathic', meaning a disease of unknown cause. Can occur spontaneously in all age groups, after pregnancy or due to poisoning with alcohol or other toxic substances. Causes chronic heart failure.

cardioplegia: a cold (4°C) clear or blood-based solution infused into the coronary arteries to stop and protect the heart in a flaccid state during surgery with the heart–lung machine. Usually contains a high concentration of potassium. At the end of the repair the heart is re-animated by restoring normal coronary blood flow.

cardiopulmonary bypass (CPB): process whereby the patient's blood is diverted away from the heart and lungs for the duration of the surgical repair. Contact of the patient's blood with the synthetic surfaces in the pump-oxygenator system elicits an inflammatory response. This limits the safe duration of blood–foreign surface interaction. The longer the procedure, the more damaging is the whole-body inflammatory response.

capillaries: billions of microscopic single-cell-thick blood channels that exchange nutrients, oxygen, carbon dioxide and metabolic by-products with the tissues of the body.

CentriMag ventricular assist device: an external magnetically levitated centrifugal blood pump widely used for temporary circulatory support. Now marketed by Thoratec for use in cardiogenic shock.

chronic heart failure: the left ventricle fails gradually but inexorably due to a number of conditions, the commonest being coronary artery disease. Causes severe breathlessness and fatigue. Conveys a high mortality rate by two years.

congenital heart disease: heart deformity that the patient is born with (e.g. atrial septal defect, ventricular septal defect, dextrocardia).

coronary artery bypass grafts: operation to bypass the narrowing of the heart's own arteries using pieces of the patient's chest-wall arteries, forearm arteries or leg veins.

coronary artery disease: gradual narrowing of the coronary arteries by atheroma. These fatty, cholesterol-based plaques are prone to rupture when they suddenly occlude the vessel, which then clots (coronary thrombosis).

CT scan: X-ray-based three-dimensional imaging of the chest and heart. By adding contrast medium the coronary arteries can be shown in detail.

deoxygenated blood: bluish blood leaving the tissues and returning to the right heart, now low in oxygen and carrying carbon dioxide to be expelled by the lungs. *See also* oxygenated blood.

diastole: relaxation and filling phase of the ventricles.

echocardiogram: non-invasive ultrasound examination of the heart chambers.

electrocautery: the electrical instrument used to cut through tissues and simultaneously coagulate blood vessels to stop bleeding.

endocarditis: bacterial infection that can destroy the heart valves.

extracorporeal membrane oxygenation (ECMO): circuit outside the body with blood pump and long-term oxygenator (lasting days) used for temporary circulatory support in acute heart failure or severe lung failure. Attached to the body by percutaneous (through the skin) cannulation of the blood vessels to the leg. Usually functions as a bridge to a longer-term pump or transplantation.

heart–lung machine: circuit outside the body to keep the patient alive while the heart is stopped for repair. Contains a mechanical blood pump and a short-term (lasting hours) complex gas exchange mechanism known as the oxygenator (artificial lung). Other pumps are used for suction of blood into the reservoir and for delivery of cardioplegia fluid to stop the heart.

HeartMate left ventricular assist device: an obsolete large pulsatile implantable pump widely used for bridge to transplant in the 1990s. The first device to be implanted on a permanent basis. Thoratec went on to produce a successful rotary blood pump for permanent use.

heart transplant: removal of the patient's diseased and failing heart, then replacement with an organ from a brain-dead donor.

heart valve replacement: removal of a diseased heart valve, then replacement with a prosthetic valve. Prosthetic valves can be biological (e.g. pig's valve) or mechanical (e.g. pyrolytic carbon tilting disc valves).

hypertension: high blood pressure that makes the heart work too hard. The level depends upon tone in the peripheral arteries. Can be very high (>200/120) and cause heart failure or stroke.

hypotension: low blood pressure (<90/60). Can be caused by blood loss or left ventricular failure. When pressure falls below 60/40 the patient is in shock and needs urgent resuscitation, and the kidneys cease to produce urine.

intra-aortic balloon pump (IABP): long, sausage-shaped balloon that's inserted into the aorta. When inflated in diastole and deflated during systole it serves to lower the resistance against which the left ventricle has to pump. Used to support the left ventricle when it's struggling to cope. Ineffective in shock states when the blood pressure or blood volume is low.

Jarvik 2000: thumb-sized rotary blood pump that is inserted into the failing heart on a long-term basis. A long-term 'off the shelf' solution for severe heart failure. The longest implant exceeds eight years.

left atrium: collecting chamber for blood returning to the heart from the lungs. The blood then passes through the mitral valve into the left ventricle. *See also* right atrium.

left ventricle: powerful, thick-walled conical chamber that pumps blood through the aortic valve and around the body. *See also* right ventricle.

left ventricular assist device (LVAD): mechanical blood pump to maintain the circulation and rest the ventricles when the heart fails catastrophically. The cannulas are inserted into the chambers of the heart. There are inexpensive temporary external devices suitable for several weeks of support in acute heart failure (e.g. CentriMag or Berlin Heart). The small, implantable but very expensive high-speed rotary blood pumps (e.g. Jarvik 2000) can be used for as long as ten years in chronic heart failure. As such, the long-term LVADs offer an off-the-shelf alternative to heart transplantation.

magnetic resonance imaging (MRI): non-invasive (without X-ray), detailed study of an organ's (e.g. heart) morphology.

metabolic derangement: consequence of poor tissue blood flow. Arteries to the muscles clamp down and the tissues produce lactic acid and other toxic metabolites.

mitral stenosis: narrowing of the mitral valve between left atrium and left ventricle caused by rheumatic fever. Flow through the valve is restricted, causing breathlessness and chronic fatigue.

myocardial infarction: death of part of the heart when a coronary artery occludes suddenly. Dead muscle is replaced by scar.

myocarditis: virus infection of the heart muscle itself, causing the heart to fail.

oxygenated blood: bright red blood saturated with oxygen and pumped around the body by the left ventricle. *See also* deoxygenated blood.

perfusionist: technician who controls the heart–lung machine and ventricular assist devices.

pericardium: fibrous sac that surrounds the heart. Can be used as patch material in the heart. Calf pericardium is used to make bioprosthetic heart valves.

pulmonary artery: large, thin-walled vessel that carries blood from the right ventricle to the lungs. It bifurcates into right and left pulmonary arteries.

pulmonary oedema: 'water on the lungs' that occurs when the left ventricle fails. Is often frothy and blood-stained.

pulmonary veins: four veins leaving the lungs to bring blood back to the heart.

reperfusion: the process whereby blood is allowed back into the coronary arteries and heart muscle following cardioplegia and cardiac arrest during surgery. The heart is re-animated and begins to beat again.

rheumatic fever: autoimmune condition triggered by a streptococcus bacterial infection that damages the heart valves and joints. Very common cause of valve disease in the pre-antibiotic era.

right atrium: collecting chamber for blood returning to the heart from the body via the veins. The blood then passes through the tricuspid valve into the right ventricle. *See also* left atrium.

right ventricle: crescent-shaped pumping chamber that propels blood through the pulmonary valve and to the lungs. *See also* left ventricle.

shock: condition when the heart cannot continue to supply sufficient blood and oxygen to the tissues. Cardiogenic shock occurs after a heart attack. Haemorrhagic shock follows profuse bleeding of two litres or more.

systole: phase of the heart cycle when the ventricles contract to expel blood.

veins: thinner-walled vessels that return blood to the heart.

vena cava: large vein entering the right atrium. The superior vena cava drains the upper part of the body; the inferior vena cava drains the lower half.